# Couples Therapy

*How to Develop a Deeper Connection with Your Partner, Improve Intimacy and Rejuvenate Your Relationship*

*Maia Daves*

# TABLE OF CONTENTS

# Introduction

We all realize that romantic bonds are hard work. Like vehicles, in order to maintain them working smoothly, they need daily maintenance. It's better to get it fixed right away to prevent more problems down the line if there is a crisis. Often, we ourselves will do some of the basic maintenance and fixes. On other occasions, we need to focus on a specialist to take a look and lend us a hand, considering our best efforts. How simply and efficiently we take those measures to fix or avoid harm to our cars is interesting. But we always stop taking steps when it comes to our relationships until the problem has become even more severe. Sadly, after a large amount of injury has also been incurred, partners often attempt pair counseling. The aim of all good therapies for spouses is to disrupt those vicious loops of miscommunication, resentment, aggression, emotional and physical abuse. But, with too many differences in the treatments required for couples, the problem persists. Which was found to be the most efficient? Most people feel that when a breakup or divorce is looming, you can just undergo relationship therapy. Sometimes, though, it is too little, too late. As soon as issues get in the way of your everyday life, relationship counseling can start. Couples counseling is a form of psychotherapy in which a therapist with psychiatric expertise dealing with couples, most commonly a Certified Marriage and Family Therapist [LMFT], helps two persons engaging in an intimate relationship obtain insight into their relationship, overcome disputes, and utilize a number of psychological strategies to increase relationship satisfaction. It is so popular to claim "relationships are complicated" that it's a cliché now. Yet it's real as well. Stress and everyday life can create disputes that appear complicated or sometimes unlikely to overcome, even though individuals get along pretty well. Relationship therapy may assist persons in progressing with their issues in these stressful circumstances, stepping through them, and be happier partners overall. We feel comforted and encouraged, our anxiety is minimized, and we learn that our attachment figures can be relied on in difficult times when our attachment figures react to our anxiety in ways that fulfill our needs. But if parents frequently respond to the frustration of a child by downplaying their feelings, ignoring their requests for support, or having the child feel stupid, the child will learn not to accept their attachment figures for support and to ignore and cope with their problems and feelings alone. When dreaming about getting into a long-term, serious relationship, passion is a top priority. In reality, 88% of Americans say that the most important factor to consider getting married is passion. We want to be cherished by our partner and fall in love with her. More than ever before, marriages are met with more strain. Couples are often dealing with relational communication difficulties and holding love intact, in addition to the long-standing stress of topics such as finances, life changes, and family dynamics. When we do not experience that kind of bond in our relationship, we turn to our spouses for warmth, reassurance, and closeness and feel hurt. Partners can get trapped in dysfunctional cycles of disconnection and start worrying with the time that they are no longer supposed to be together. "Couples counseling" and "couples therapy" are usually considered as the same thing. On a scientific basis, there is little distinction between them. The other way in which whether the session is called matters is a valid one; in certain places, you may get a separate "therapy" qualification or license to practice that is tougher to receive than the "counseling" qualification or license to practice. This form of relationship with a trained therapist presents partners with an ability to work on their most complicated or socially demand-ing concerns, whether you name it partners therapy or couples counseling. These topics may vary from basic difficulties of un-derstanding or serious disputes to difficulties of drug misuse and psychiatric conditions. Although counseling for partners may be a wonderful way to bond with your spouse or mend the gaps between you, without having a therapist, there are also ways to ensure sure you maintain the flame is going and the relationship safe. There are several tools out there that rely on couples counseling ideas or tests. It is never too late to start add-ing a little more time into your relationship (or too early). Choose one or two of the practices and tasks listed below to practice with your spouse if you would like to strengthen your relationship. If there is a fully efficient tool out there for stable, safe marriages, certainly somebody might have picked it up and marketed it by now, right? We will have to do with what we have before we can discover the 100 percent performance promised blueprint for a perfect relationship, building our communication skills, connecting efficiently, participating in events the strengthen our bond, and utilizing couples counsel-ing to resolve some of the major issues. In order to develop a stable relationship and ward off divorce or breakup, there is no "best" behavior for partners because each partner would have their own best practice. For certain couples, it might be joining together in a common sport, such as riding a horse, playing a classic series, or playing guitar together. For some, when gazing up at the night sky, over coffee in the morning, or lying-in bed at night, it could be the prolonged chats they sometimes have.

# Chapter 1: Couples Therapy Basics

We all realize that romantic bonds are hard work. Like vehicles, in order to maintain them working smoothly, they need daily maintenance. It's better to get it fixed right away to prevent more problems down the line if there is a crisis. Often, we ourselves will do some of the basic maintenance and fixes. On other occasions, we need to focus on a specialist to take a look and lend us a hand, considering our best efforts. How simply and efficiently we take those measures to fix or avoid harm to our cars is interesting. But we always stop taking steps when it comes to our relationships until the problem has become even more severe. Sadly, after a large amount of injury has also been incurred, partners often attempt pair counseling. Maladaptive marital behaviors have been ingrained, the relational bonding amongst spouses has been significantly compromised, and due to unresolved past disputes, there is a high degree of distrust. Some think of counseling as a means to push their spouse to improve when they are "the question." In treating a broad spectrum of relationship difficulties, often, persons are not conscious of the advantages of pair counseling. They may not realize how critical it can be to improve the overall happiness of relationships that influence individual mental wellbeing.

# What Works in Couples Therapy?

**What are the two most effective couples' therapies, and what do they share?**

The aim of all good therapies for spouses is to disrupt those vicious loops of miscommunication, resentment, aggression, emotional and physical abuse. But, with too many differences in the treatments required for couples, the problem persists. Which was found to be the most efficient?

## Two Most Effective Couple Approaches

Two methods have developed as what researchers call "evidence-based" treatments for partners following detailed studies on pair counseling.

- Integrative Behavioral Pair Counseling or IBCT is considered the first method.

- The second option is Counselling for Emotion-Focused Partners or EFCT.

### Different Roots but the Same Tree?

Both of these two tactics are equally efficient. They rise, however, from very distinct philosophical and theoretical origins on the cause and cure for the problems of couples. Their ideologies couldn't be far separated in many aspects. IBCT is based on the concepts of action and perception. In humanistic and existential philosophies, on the other side, EFCT has its origins. IBCT sees couples as engaging in transactions of quid pro quo behavior, stressing incentives, and aversions in sequences framed by the expectations of couples regarding each other. Based on their histories of the previous family of origin bonds or lack thereof, EFCT considers couples as reacting to each other. Yet, when confidence and cooperation grow, all methods target the same worsening vicious loops between partners and aim to reverse them by growing recognition and insecurity and then modifying other alarming trends.

## Things to Know About Couples Therapy

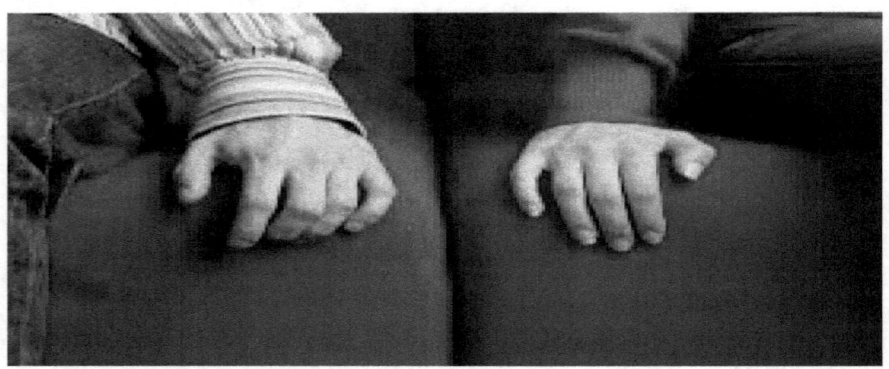

We all realize that romantic bonds are hard work. Like vehicles, in order to maintain them working smoothly, they need daily maintenance. It's better to get it fixed right away to prevent more problems down the line if there is a crisis. Often, we ourselves will do some of the basic maintenance and fixes. On other occasions, we need to focus on a specialist to take a look and lend us a hand, considering our best efforts. How simply and efficiently we take those measures to fix or avoid harm to our cars is interesting. But we always stop taking steps when it comes to our relationships until the problem has become even more severe. Sadly, after a large amount of injury has also been incurred, partners often attempt pair counseling. Maladaptive marital behaviors have been ingrained, the relational bonding amongst spouses has been significantly compromised, and due to unresolved past disputes, there is a high degree of distrust. The list might continue on. Studies reveal that couples take over six years before finding therapy with partners, and the typical spouse remains miserable. This is not to say that partner therapy would not be beneficial in solving certain long-standing concerns. It will, though, be a far more complicated and time-consuming effort, with a lot of focus and effort on both sides. Misconceptions of what counseling for partners is and its intent may also discourage spouses from finding treatment early on. You may believe that counseling for partners is mainly designed for really severe conditions that impact a relationship, like an infidelity or addiction. Before making the decision to terminate the relationship, some might perceive it as a last-ditch attempt. Some think of counseling as a means to push their spouse to improve when they are "the question." In treating a broad spectrum of relationship difficulties, often, persons are not conscious of the advantages of pair counseling. They may not realize how critical it can be to improve the overall happiness of relationships that influence individual mental wellbeing.

## What Is Couples Therapy?

Couples counseling is a form of psychotherapy in which a therapist with psychiatric expertise dealing with couples, most commonly a Certified Marriage and Family Therapist [LMFT], helps two persons engaging in an intimate relationship obtain insight into their relationship, overcome disputes, and utilize a number of psychological strategies to increase relationship satisfaction. While the practice of pair therapy can differ based on the theoretical orientation of the therapist, both pair therapy appears to contain the following basic elements:

- An emphasis on a single issue (i.e., sexual problems, addiction to the Internet, jealousy)

- Strong intervention by the client in the care of the relationship itself, rather than independently by each client.

- Solution-focused, early on in care, change-oriented approaches.

- A consistent set of priorities regarding recovery

Couple counseling typically begins with the typical interview questions about the relationship's past as well as some discussion of the families of birth, beliefs, and cultural context of each spouse. If required, the psychiatrist would use the initial sessions for crisis intervention. The psychiatrist of the pair would also help the pair in determining the dilemma that would be the subject of therapy, setting expectations for therapy, and preparing a treatment structure. The psychiatrist can help the pair obtain insight into the relationship complexities that sustain the issue during the recovery process, thus making both parties recognize both of their positions in the unhealthy interactions. This will allow them to modify the way they and each other view the relationship. While it is necessary to obtain understanding, another vital element of pair counseling actually includes altering habits and methods of communicating with each other. Couple practitioners also delegate homework to couples to apply their day-to-day experiences to the techniques they have gained in counseling. Many couples will get away from the counseling of couples who have gained insight into intimacy dynamics, improved emotional expression, and built the skills required to interact and manage issues more efficiently for their spouses.

## Who Is It For?

For any type of relationship, if couples are heterosexual, lesbian, mixed-race, female, elderly, single, engaged, or married, couple counseling is useful. For starters, before getting married, a newly engaged pair may find premarital therapy an invaluable opportunity to discuss relationship expectations. Another pair, 25 years married, will find that counseling for partners is an efficient way for them to restore a sense of excitement and intimacy in their relationship. Couple counseling may address a current conflict, avoid issues from exacerbating or even offer a "check-up" for a satisfying couple that is facing an adjustment phase or heightened tension. In couples counseling, common topics of focus include financial concerns, parenting, ethnicity, infidelity, in-laws, chronic health conditions, infertility, gambling, drug use, emotional distance, and regular confrontation.

# Online Couples Therapy vs. Face-to-Face Couples Therapy

Today, internet counseling has not only revolutionized the way clinicians have practiced counseling with families. It has also been encouraged to reduce the shame of seeking counseling for spouses and to make therapy for spouses more available and inexpensive. Couple counseling has many benefits with a text-based setting that is not offered with conventional face-to-face counseling for partners. Second, given the freedom to share as frequently as one wishes, without any interruptions, there is the potential for a lot of contacts. Before composing them and expressing them in the counseling session, the capacity to process one's own feelings results in couples interacting more appropriately and more consistently. The format often allows each participant the time and space to process the answer of their participant and focus on what they have "seen" instead of only reflecting on what they are about to suggest. Some people often feel that by writing, they can further describe themselves. This will then allow for emotional expression that is more accessible and truthful, increasing intimacy between partners. This pair therapy approach is often suitable for couples when either or both spouses often move or have very busy lives and are unable to arrange a time to engage together in pair therapy sessions. Couples with kids that have trouble accessing childcare will still profit immensely from the convenience of counseling for online couples. Often, couples counseling is not commonly provided by many health care policies, and with just a few treatments, it may be very pricey for many people. Therapy for online lovers will be a far more accessible option. This format makes for more fruitful discussions, resulting in much faster treatment change.

**Everything You Need to Know About Relationship**

### Counseling

It is so popular to claim "relationships are complicated" that it's a cliché now. Yet it's real as well. Stress and everyday life can create disputes that appear complicated or sometimes unlikely to overcome, even though individuals get along pretty well. Relationship therapy may assist persons to progress with their issues in these stressful circumstances, step through them, and be happier partners overall.

## When to Seek Relationship Therapy

Most people feel that when a breakup or divorce is looming, you can just undergo relationship therapy. Sometimes, though, it is too little, too late. As soon as issues get in the way of your everyday life, relationship counseling can start. Here are some indications that a consultation could help you:

- You have difficulties voicing your thoughts to one another.

- You have one or more unsolvable disputes.

- In the interactions, there is withdrawal, critique, or scorn.

- A traumatic incident has rattled the normal existence.

- You have difficulty with shared choices.

- You also encountered unfaithfulness, addiction, or suspected violence.

- Try to build a better partnership.

Bear in mind that there are no bad motives for finding marital therapy. In order to avoid severe issues from arising, certain couples initiate counseling as soon as they are married, often without apparent problems. Counselors will help you become a stronger communicator, develop good skills in relationships, and increase the happiness of your family. Bear in mind that before receiving care, the typical couple waits for six years. This is a lot of time to let concerns fester; troubled marriages are hard to save at this stage. Therefore, finding issues early and pursuing the treatment as soon as possible is critical. Relationship issues are not confined to sexual ones, although it is the most common explanation people consult for relationship counseling.

## Premarital Counseling

Premarital counseling is a form of therapy for marriages that helps brace partners for a long-term marriage. This style of therapy focuses on helping partners build a good and stable pre-marriage relationship and finding any possible issues that could contribute to issues down the path. Many of the intimacy problems that could be discussed through premarital therapy include:

- Beliefs and convictions

- Responsibilities and roles

- Sex and love

- Finance

- Interaction

- To choose whether or not to have children.

- Parental decisions

- Family support

This method of relationship therapy will be a successful means of setting reasonable goals and improving positive communication habits that can provide a good start to a marriage.

## How to Find a Relationship Therapist

There are a variety of practitioners, including qualified psychologists, a certified marriage and family therapists, licensed counselors, and accredited qualified social workers, who are willing to provide relationship counseling. Note that you may not need to be married to benefit from marital therapy, even if the title says "marriage". While turning on the internet is the first instinct of most people when searching for a psychiatrist, a more reliable place to proceed is to ask for referrals from people you meet. There are potentially hundreds of skilled therapists if you reside in a metropolitan city, and making the decision can be daunting. If someone you meet has met with a psychiatrist professionally, there's a fair possibility they may also work with you. Take advantage of the complimentary consultation for prospective new customers that many therapists provide. This is a perfect time to see if the desires, personality, and budget are balanced by the specific advisor. In several meaningful ways, therapist-client relationships will influence your life, and you can choose carefully.

## Online Relationship Counseling

Online counseling may be a perfect alternative if conventional face-to-face therapy won't fit with you and your partner. There are a number of reasons why you might want to try therapy online:

- You move to various places with your partner. This could refer to persons or others who are divorced and contemplating a permanent separation or are in long-distance relationships. And if they reside separately, online counseling programs offer all couples the opportunity to join.

- You fly mostly for jobs. Online options enable individuals to profit from therapy, regardless of how full their life is or where they are situated in the world.

- Conventional care is not comfortable for you or your partner. For certain individuals, face-to-face counseling may be demanding, awkward, or even anxiety-provoking. Web-based solutions will make things more available for relationship therapy.

Tools such as internet chats, video sessions, and voice calls are utilized for internet relationship therapy providers where partners can speak to each other and their psychiatrist. You and your wife may work through counseling to build expectations you will like to accomplish, which could involve resolving communication-related issues, conflicts, or infidelity.

## What to Expect

Your past and the challenges you are there to address will rely on the first few sessions. Be prepared before the new one to address questions regarding the engagement, your partners, your upbringing, and engagement memories. Your doctor may want to spend more time talking to everybody together, then individually with each participant. The direction your recovery is going to proceed relies on the counselor's personality and the clinical technique they are utilizing. Emotionally oriented counseling, or EFT, is the most practiced style of relationship counseling. EFT is based on the philosophy of relationships and connections, which seeks to promote healthy interdependence between the pair or family members. Imago counseling and the Gottman technique provide several forms of intimacy counseling. Tell the psychologist the strategy they are skilled in and the one they feel is more suitable for the case.

# Chapter 2: Traditional and Contemporary Methods for Couples Therapy

Since we all require a human bond, why don't we love the same thing? Research has found about 35-40 percent of individuals claim they feel uncomfortable in their adult relationships over many years and many environments, whereas 60-65 percent experience stable, caring, and rewarding relationships. In part, how safe or vulnerable we are with our intimate partners depends on how, at a young age, we interact with our parents. We have turned to our parents or guardians from the day we were born for affection, warmth, and protection, especially in times of distress. We name them 'attachment statistics' for this purpose. We feel comforted and encouraged, our anxiety is minimized, and we learn that our attachment figures can be relied on in difficult times when our attachment figures react to our anxiety in ways that fulfill our needs. But if parents frequently respond to the frustration of a child by downplaying their feelings, ignoring their requests for support, or having the child feel stupid, the child will learn not to accept their attachment figures for support and to ignore and cope with their problems and feelings alone.

# The Benefits of Couples Therapy While Separated

When dreaming about getting into a long-term, serious relationship, passion is a top priority. In reality, 88% of Americans say that the most important factor to consider getting married is passion. We want to be cherished by our partner and fall in love with her. More than ever before, marriages are met with more strain. Couples are often dealing with relational communication difficulties and holding love intact, in addition to the long-standing stress of topics such as finances, life changes, and family dynamics. When we do not experience that kind of bond in our relationship, we turn to our spouses for warmth, reassurance, and closeness and feel hurt. Partners can get trapped in dysfunctional cycles of disconnection and start worrying with the time that they are no longer supposed to be together.

## Separation Before Divorce

They can begin to believe that things are finished and will not be healed or fixed when couples realize their relationship is in distress. However, contemplating a separation when discerning what step to take next in their relationship can be fair and advantageous for couples.

## Choosing a Trial Separation

For couples struggling in their relationship but uncertain

whether divorce is the necessary next action to take, a trial separation may be an alternative. They may prefer to live in different places while couples are not getting along when they continue to navigate through problems, they face inside themselves and within the relationship. Any individuals deem a breakup from the trial to be a "one foot out of the house" shift and a path to divorce or the eventual end to the relationship. Each pair is distinct, and there are a number of explanations for entering a split from the prosecution. Among these families, divorce is not unavoidable. In fact, during this period, when couples are living separately and may be relatively separated from the dysfunctional habits, they encountered with each other when living together, relationship therapy may be of considerable help.

## How to Suggest Counseling

You might ask whether it is still a reasonable opportunity to speak to your spouse about therapy and see whether they are going to go with you. The truth is, the safest way to inquire is whether you feel that your partnership might be helped by therapy. If you are divorced but still think that therapy for partners might be helpful to your relationship, the benefit of asking your spouse to engage in therapy might outweigh the harm. So, how can you question a buddy of yours? Bear in mind that anxiety most commonly prevents individuals from joining the process of therapy. In the meantime, you or your partner might be scared to feel more emotional distress or even to be seen as the "evil guy" or the "broken one." Take time to focus on your own thoughts regarding the therapy phase and whether your spouse might be fearful of getting underway. Enable both of you room to chat freely about the problems and, if necessary, make an attempt to collectively study counselors to identify someone with which you both believe you might be happy.

**Finding Your Counselor**

There are several psychologists and other professionals who claim they deal with families, but in this advanced work, they are not necessarily well qualified, so it would be beneficial for you to do more homework before finding a psychologist for your particular case. The counselor's training directly linked to marriage and relationship counseling is one key consideration to remember. You may want to trust that the psychologist you chose will appreciate the fragile condition of the relationship when helping you settle down and explore the waters of healing and rebuilding relationships. Don't be afraid to ask a few other counselors to ask questions regarding the programs they provide in your city.

**Questions to Ask a Counselor:**

- Can you feel relaxed dealing with disconnected couples?

- Do you rely only on marriage and relationships?

- How long have you been dealing with couples?

- When we continue therapy for you, what do we expect?

- What's your relationship therapy training?

Taking time to raise questions such as these from a psychologist will allow you to obtain a greater understanding of their advanced skills, their expertise in communicating with families, and how they can support you and your spouse through this stressful time.

**Benefits of Separation Counseling**

There are tons of advantages a couple will encounter when apart in counseling. Much like it does with families who still work together. Counseling will allow you to recognize the trends that have arisen that have contributed to this location, how to achieve clarity, and learn from the experience in order not to replicate old habits. In fact, separation may help to de-escalate the tension for high-conflict partners, enabling marital therapy to act as a secure place to start processing what is occurring in their dynamics. Counseling will also serve to add understanding and harmony to complex relationship decision-making. Relationship therapy can give you any of the following if you and your spouse are currently separated:

- Guidance for maintaining a successful transfer back home.

- Opportunities for de-escalating existing conflicts.

- Professional support to heal the partnership and fix it.

- Reconciliation in order to create a solid, stable relationship.

- A healthy environment for seeing and listening to how the dispute impacts each partner.

- A sensation of the possibility of reconnection.

- A place to process uncomfortable feelings about the actions to take next.

- Time to learn what might have contributed to a disconnection.

- Trusted advice for navigating tough relationship choices.

There is no question that spouses who are separating or entering a position of separation are in pain. Each partner's feelings are likely to be strong, whereas expectations for progress and development are poor. Couple therapy will provide you and your wife with ample room and time to decide which measures to follow in your relationship next. With the aid of a qualified marital counselor, healing and repair can be an alternative.

## What Is 'Attachment' and How c It Affect Relationships?

Since we all require a human bond, why don't we love the same thing? Research has found about 35-40 percent of individuals claim they feel uncomfortable in their adult relationships over many years and many environments, whereas 60-65 percent experience stable, caring, and rewarding relationships. In part,

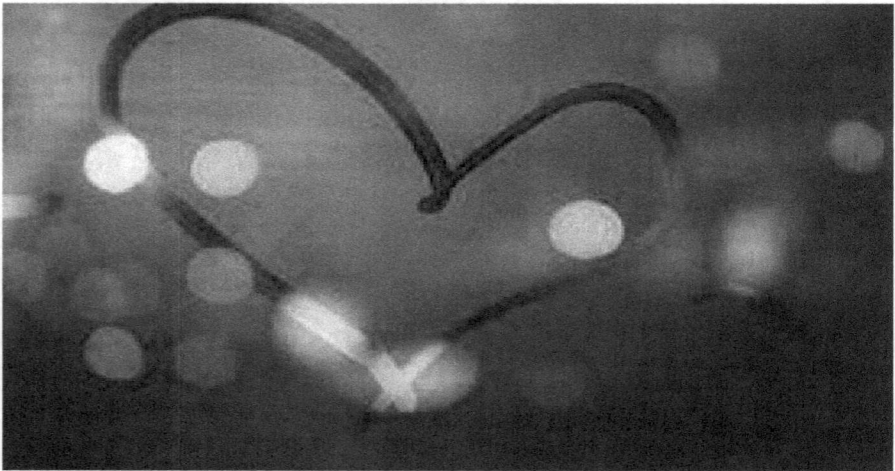

how safe or vulnerable we are with our intimate partners depends on how, at a young age, we interact with our parents.

We have turned to our parents or guardians from the day we were born for affection, warmth, and protection, especially in times of distress. We name them 'attachment statistics' for this purpose. We feel comforted and encouraged, our anxiety is minimized, and we learn that our attachment figures can be relied on in difficult times when our attachment figures react to our anxiety in ways that fulfill our needs. But if parents frequently respond to the frustration of a child by downplaying their feelings, ignoring their requests for support, or having the child feel stupid, the child will learn not to accept their attachment figures for support and to ignore and cope with their problems and feelings alone. These methods of downplaying are called "deactivating attachment methods." For some, parents respond to the frustration of a child by being inconsistent with the help they offer or by not offering the correct form of care. Perhaps often they understand the pain of their infant; other times, they do not understand the pain or dwell on how the distress made them sound rather than helping the child control their emotions. Or maybe certain parents have help, but that's not what the kid wants. A child might require motivation to cope with a problem, for instance, but the adult tends to be sympathetic and accepts that the child may not manage the problem. Daily exposure to these forms of parental interactions implies that, particularly while overwhelmed, these kids will feel extreme concern and make a lot of effort to stay really close to their attachment figures. The so-called "hyper-activating techniques" are these techniques of growing interest and pursuing unnecessary closeness. It has been discovered that our attachment styles influence the way we initiate, sustain, and terminate relationships. Unsurprisingly, in intimate relationships, those who have a stable attachment style seem to do better. They report the highest happiness in marriages, appear to cope with disagreements by participating in positive activities, listening to the point of view of their spouse, and doing a decent job of controlling their emotions. These individuals often assist their spouses efficiently, both in periods of depression and prosperity. These individuals appear to engage more confidently with prospective partners when it comes to relationship initiation. They often participate in an acceptable degree of transparency regarding themselves. They appear to feel less unpleasant feelings as they split up from a relationship, indulge in less partner-blaming, and are more inclined to look to individuals for help. They often display a stronger inclination than other insecurely attached individuals to acknowledge the failure and start dating earlier. Many who feel uncertainty of commitment appear to report less satisfaction with the relationship. Those high on attachment anxiety appear to indulge in confrontation and do it in a detrimental fashion that entails utilizing critique, blame, and seeking to make the other feel bad. They may be over-helpful as they indulge in care, and then the care may come off as smothering or overbearing. These individuals can come off as very open and partner in terms of establishing relationships but can over-disclose very early in the relationship and attempt to continue the relationship at a rapid rate. They will find it impossible to let go when it comes to break-ups, feel a high degree of pain, and use various strategies to get back into their partner.

## What are the Attachment Styles?

These techniques, together with the opinions and emotions of people regarding relationships, shape the base of the attachment style of an individual in adulthood. Our own type of attachment is the product of how we assess two variables: attachment anxiety and avoidance of attachment. Attachment anxiety differs from low to severe, with individuals high on attachment anxiety having a high need for acceptance, an overwhelming need to remain physically and emotionally connected to others, especially intimate partners, and difficulty in relationships managing their pain and emotions. Attachment avoidance often differs from low to large, with persons large on attachment avoidance showing a mistrust of others, personal and emotionally near anxiety, extreme self-reliance, and a willingness to ignore their concerns and feelings. Individuals that score low on both attachment distress and avoidance provide a healthy attachment. They trust people, are relaxed with expressing feelings and getting close to others, and prefer not to downplay their discomfort or exaggerate it. They still feel secure in addressing difficulties and life stressors and looking to others for assistance.

## Can they change over time?

Throughout life, our relationship patterns are considered to be moderately healthy, while certain individuals manage to shift from an unstable relationship to a comfortable attachment type. It doesn't just happen; however, it requires a lot of time. Research shows that while attachment styles can become more difficult to alter as we mature, life events and interactions that contradict our pre-existing relationship values can contribute to improvements in our style of attachment. In order to minimize attachment insecurity, getting married, and creating mutual aspirations that affirm affection and loyalty to another have been identified. But incidents that are perceived as challenges to one's partnership or the lack of interaction, such as spouse rejection, may raise the vulnerability of attachment.

## How do they Affect our Romantic Relationships?

It has been discovered that our attachment styles influence the way we initiate, sustain, and terminate relationships. Unsurprisingly, in intimate relationships, those who have a stable attachment style seem to do better. They report the highest happiness in marriages, appear to cope with disagreements by participating in positive activities, listening to the point of view of their spouse, and doing a decent job of controlling their emotions. These individuals often assist their spouses efficiently, both in periods of depression and prosperity. These individuals appear to engage more confidently with prospective partners when it comes to relationship initiation. They often participate in an acceptable degree of transparency regarding themselves. They appear to feel less unpleasant feelings as they split up from a relationship, indulge in less partner-blaming, and are more inclined to look to individuals for help. They often display a stronger inclination than other insecurely attached individuals to acknowledge the failure and start dating earlier. Many who feel uncertainty of commitment appear to report less satisfaction with the relationship. Those high on attachment anxiety appear to indulge in confrontation and do it in a detrimental fashion that entails utilizing critique, blame, and seeking to make the other feel bad. They may be over-helpful as they indulge in care, and then the care may come off as smothering or overbearing. These individuals can come off as very open and partner in terms of establishing relationships but can over-disclose very early in the relationship and attempt to continue the relationship at a rapid rate. They will find it impossible to let go when it comes to break-ups, feel a high degree of pain, and use various strategies to get back into their partner. By separating from their spouses, physically breaking down, and declining to address problems as they emerge, people high in attachment avoidance prefer to prevent confrontation. They still find it challenging to offer assistance, and they do so in a withdrawing and uninvolved manner if they are obligated to assist their spouse. In terms of beginning relationships, in the early stages of a relationship, those high on commitment avoidance feel emotionally uninvolved and distant and may appear to portray an over-inflated self-image. Individuals heavy on avoidance prefer to report low levels of anxiety in terms of a relationship breakdown and do not seek past spouses. They prefer to go about things in a round-a-bout fashion if a split were to arise, to prevent directly stating that they expect the partnership to stop, to prevent tension and awkward conversations. In moments of tension, the variations in how firmly and insecurely connected individuals' function in their relationships are more apparent. Several researchers have found that tension raises the likelihood of detrimental consequences for vulnerable individuals: decreases in partner fulfillment and raises in confrontation attitudes that are harmful.

## How Can you Boost your Security?

Growing someone's sense of safety can be accomplished in a range of ways. One includes exposure to words or photographs that inspire feelings of affection, warmth, and attachment (such as showing people a picture of a mother carrying an infant, hugging a couple, or terms such as "hug" and "affection"). One includes exposure to terms or photos. Another is to help them remember past events where they were comforted by a person. Another line of study has explored how spouses should better help each other to either reduce or mitigate the vulnerability of attachments. Preliminary literature shows that it is a positive idea for those high in attachment distress to help individuals feel secure and improve their self-confidence. Not being as hostile and dismissive during disputes or when coping with interpersonal difficulties might be the safest approach for anyone high in attachment avoidance. A clinical technique called Emotionally Oriented Couples Therapy (EFCT) has been introduced within the area of relationship counseling to resolve the detrimental effects of intimacy insecurity in intimate couples and has been shown to be successful. EFCT works on breaking loops of detrimental relationship interactions by getting all members of the pair to cope with the worries by expectations of each other's attachment, such as abandonment and rejection. Couples then explore from a therapist how to more accurately express their attachment criteria to each other for affection, warmth, and stability. For others, the search for healthy and caring interpersonal interaction is a real struggle, yet meaningful impressions of potential interactions have the ability to shift individuals from a position of vulnerability to one where affection, approval, and warmth can be found.

## Integrative Behavioral Couples Therapy- Creating Acceptance and Change

### The Essence of IBCT

Initially, behavioral pair therapy experts developed pair treatments focused strictly on theories of mutual adjustment in the behaviors of problem partners. They found, though, that not all partners were responsive to compromise and obedience when they approached counseling, and it eroded with time even as improvement occurred. Thus, the researchers applied what they considered an aspect of approval of the therapy.

## Strategies to Build Acceptance

The following is a collection of methods focused on acceptance and responsiveness employed by the IBCT strategy.

### Empathic joining around the problem

One of the first acceptance phases is empathic joining around the problem. Therapists stress each partner's discomfort and sincere efforts without accusation. By casting the couple's problem as flowing from an honest discrepancy that inevitably contributed to polarization and ultimately separation, the formulation itself became a key mechanism for this marriage. In their reflections, clinicians stress using "sensitive" rather than "strong" emotion definitions, and they urge couples to do the same.

### Encouraging unified detachment

Encouraging unified separation or establishing an "it" question results from empathic joining. "Instead of accusing each other, the dilemma is externalized, and a metaphorical name is sometimes offered as" it. "An example is" Two porcupines attempting to dance.

## Highlighting the positive features of negative behavior

Highlighting the beneficial features of undesirable conduct specifically contributes to the dialectical or polar essence of the problem of the pair. In relationships, complementary disparities typically establish equilibrium. Such discrepancies may also become a good part of the relationship between the pair and something for them to be proud of and feel connected to.

## Role-playing negative behavior

A widely employed strategy in IBCT is harmful role-playing actions during the counseling session. This includes making the pair carry out the pattern in session through role play. The principle is to assist the pair to do what they had been doing instinctually on intent. By desensitizing the couple to the sequence, the goal is for the relationship to decrease their arousal levels. Lowered arousal allows problem-solving simpler in the face of the question.

## Faked incidents of negative behavior

Faked instances of unpleasant behavior at home or "faking evil" are explained by desensitizing spouses of negative behavior as a means of encouraging empathy. Partners are asked to select battles purposely or do or not do something purposely that causes tension while they would not usually do this. The faker is then challenged to watch the mechanism of the vicious loop, to help monitor and benefit from the polarization and suffering. It has important impacts on the de-escalation of wars and the improvement of the formulation of how disputes take shape.

## Preparation for backsliding

Backsliding training is an inoculation that is used when improvement is made. It is ultimately relapse avoidance that inoculates the couples from both believing that the epidemic has begun again and seeking to resolve it again using first-order remedies of the same old vicious circle.

## Emotional acceptance through greater self-care

There are two elements to emotional acceptance through greater self-care. The first is to make each partner continue to do stuff to feel better and to practice self-soothing for him or herself. The second is to understand during debates, polarization cycles, and other situations of unpleasant actions to do self-care. Further trend changes may then be rendered as greater acceptance is created.

# Emotionally Focused Therapy

Emotionally Focused Therapy was established in the 1980s as a short-term and organized alternative to pair counseling, typically lasting from eight to twenty sessions. It reflects on unhealthy contact habits and affection as an attachment bond and is focused on research.

## Emotion-Focused Couples Therapy (EFCT)

In three stages, EFCT is delivered. There is an interview prior to Step One to ensure that the pair is appropriate for therapy, requiring eight activities. This involves the unavoidable phases of forming a joint relationship, including bonds, priorities, and responsibilities, and determining the reaction of the pair to the therapist. The therapist often examines in this phase how open and sensitive the pair is to the big EFCT system that their issues exist in their bonding experience and their relational means of protecting against more wound and failure insecurity.

## Phase One

The clinical alliance, as with other therapies, is partially based on the consumers buying into the system and rationales for therapy. Clinicians support couples in the first stage to see that their concern is not one another, but actually the problematic relationship pattern, which obscures their actual needs. The need is to feel relaxed and closer to each other. This is a basic assumption of the EFCT method. EFCT aims to build more stable relationship ties between partners in line with the relationship principle, shift vulnerability expectations, and build a practical, safe haven and secure foundation for the partners to succeed. Again, it is important to tailor this frame to the pair in order to step on in care.

**Phase Two**

Step Two contains what is referred to as reforming the relationships of the pair to achieve a stable bond of connection and attachment. This includes having the candidate ease his or her approach by voicing the emotions on the soft side that are behind moaning and frustration. This usage of the word "soft feelings" is curiously the same as in the IBCT method. The psychiatrist, on the other hand, allows the withdrawer to participate. This requires sharing their own pains, worries, and soothing needs. Partners may obtain supportive, calming responses when the attachment is healthy while staying accessible and sensitive to the emotional needs of each other. A new form of reacting unfolds and substitutes the demand-withdrawal trend through this method.

**Phase Three**

Finally, gains are sustained in Phase Three, and, utilizing the recently learned interactive pattern, long-standing issues are resolved. There is also a discussion of past causes of the demand-withdraw loop. The advisor then serves as a facilitator in constructive problem-solving and as a framework consultant. There are obviously some respects in which the EFCT method resembles the IBCT method in prohibiting the vicious demand-withdraw loops of partners, but the outcomes are obtained through somewhat different assumptions. Of necessity, each strategy stresses empathy and bonding to strengthen the therapeutic relationship and encourages partners to buy into the strategy frames and rationales. Each prevents the same traditional vicious loops from various roots, but in the same ways in general.

**Which One Is Best?**

We are still inclined, of course, to question which one of these two is best. The response turns once more to the definition of the suit. With varying life experiences and values, every couple comes to counseling because the solution that makes the most sense to them, or suits them better, is the one that is likely to perform the better.

**Attachment Theory**

Usually, "attachment" between individuals offers a secure harbor. An escape from the environment and a way to achieve warmth, protection, and a tension shield. The attachment also creates a stable basis, enabling you to feel secure when discovering the environment and studying new details. In infancy, the development starts with a primary caretaker, such as a relative. Those early, established trends lead through to adulthood. In an infant, an "unavailable caretaker" induces distress comparable to an "unavailable spouse" that triggers distress in an adult. Attachment theory offers a "route chart" for the emotionally oriented therapist to the drama of pain, thoughts, and desires of couples. EFT is used in private practice, academic teaching centers, and hospital facilities by several different kinds of couples. For numerous ethnic communities around the globe, it is also very helpful. The troubled couples who can benefit from EFT include, among other disorders, those where either or both spouses suffer from depression, alcoholism, traumatic stress disorder disorders, and chronic disease. For couples struggling with infidelity or even more stressful events, both recent and past, the EFT has proved to be an effective approach. Attachment theory and EFT frequently interact with neuroscience. The importance of healthy attachment is shown by more recent MRI research. Our attachments are solid, and they are encoded by our brains as "protection." Any perceived difference or divergence in our near relationships is viewed as a risk, according to EFT studies. Losing a loved one's link undermines our sense of wellbeing. In a region of our brain named the amygdala, also known as the anxiety core, "Primal anxiety" ensues and sets off an alarm.

## Fight or Flight Mode

It activates our fight-or-flight reaction until the amygdala is triggered. The amygdala is relaxed, as incoming knowledge is well understood. However, it raises the level of fear of the brain as soon as the amygdala experiences threatening or unknown knowledge and centers the concentration of the mind on the immediate circumstance. Individuals move through a state of self-preservation, sometimes doing as they did in infancy to live or cope. This explains why in our intimate relationships, in the same repeating and toxic habits from our formative years, we are stimulated as adults. These involuntary, counter-productive responses may help to unwind the EFT.

## Healthy Dependency

EFT presents a vocabulary for safe commitment dependence and looks at crucial steps and moments that describe a partnership of adult affection. The model's primary purpose is to broaden and re-organize the couple's relational reactions. Existing, destructive trends such as "pursue-withdraw" or "criticize-defend" are substituted by new loops of bonding encounters. These fresh, constructive processes then become self-reinforcing and generate lasting improvement. For both spouses, the relationship becomes a sanctuary and a healing setting.

## Creating a secure bond

The method decreases the tension between partners, thus establishing a more stable romantic relationship. From the point of insecurity, partners learn to convey intense, fundamental feelings and appeal for their needs to be fulfilled. "Partners tend to perceive unwelcome actions (i.e., shutting down or angry escalations) as" disconnection protests. "Spouses grow to be physically open, empathic, and active with each other, reinforcing the bond of attachment and safe refuge between them. As a behavioral model, EFT has several benefits. Second, a detailed analysis confirms it. Second, it is constructive with customers and supportive of them. Instead of the spouses themselves (or the partners), it transfers the responsibility for the few issues to the unpleasant patterns between them. The process of change has been mapped into a well-specified process composed of nine phases and three events of change that can direct and monitor success for the therapist. An EFT qualified psychiatrist will be a wise decision if you are searching for support with a distressed relationship.

# Chapter 3: Couples Therapy with Imago, Sensate-Focus and Partner Yoga

Imago therapy is a special form of relationship therapy intended to help heal and create strategies for tension within relationships. The word imago is Latin for "picture" and applies to an "unconscious picture of familiar affection" inside imago relationship therapy. The notion of imago as a picture of familiar love implies that we understand much about love and about ourselves through our early relationships. We build a sense of an identity linked to love through these early encounters, such as what love is and what we need to do in order to encounter love from others and feel secure. Sex therapy is a type of psychotherapy that is intended to support sexual issues to be resolved by people and couples. Female therapy, not hands-on therapy, is conversation therapy. Everyone in the audience is completely dressed in a sex therapy session, although there is no contact. In order to rule out medical causes of sexual issues, sex therapists can suggest obtaining physical examinations. For those who are focusing on sexual issues that may not have a partner to experiment with, they can even promote the usage of sexual surrogates. Sex practitioners, though, should not conduct medical tests on their clients or have intimate experiences with them. Only the act of checking out your partner's yoga class for a couple will make you become happier with your relationship. Studies have found that partners who participate together in challenging new experiences will feel an improvement in both the strength of the relationship and emotional appeal. Furthermore, the intimacy and mutual posing in the yoga of couples will help to refresh and restore a bond. It helps couples to have fun whilst calming down, enjoying quality time, and having a positive relationship while gaining different skills together. Yoga often generates consciousness; which research has related to better experiences. The research reported that in a positive correlation between enhanced mindfulness, described as accessible attention to and mindfulness of the present moment, and greater satisfaction with relationships. It will reinvigorate the bond by being in the present moment while concentrating on breathing and posing with the partner, helping you both feel satisfied with your relationship.

## Imago Therapy for Relationships

Imago therapy is a special form of relationship therapy intended to help heal and create strategies for tension within relationships. The word imago is Latin for "picture" and applies to an "unconscious picture of familiar affection" inside imago relationship therapy. In the late 1970s, two therapists who had undergone divorce in their relationship histories founded

Imago relationship therapy. They wanted to draw on their perspectives to study and create an evidence-based therapy model that will help promote healing and development in committed relationships after exploring reliable and evidence-based support for recognizing relationship complexities and having relatively little in the way of supportive services.

## Imago and Relationships

The notion of imago as a picture of familiar love implies that we understand much about love and about ourselves through our early relationships. We build a sense of an identity linked to love through these early encounters, such as what love is and what we need to do in order to encounter love from others and feel secure. In our early marriages, we start to build a sense of self-worth dependent on how we are viewed by significant people in our lives. We start forming habits of attachment and continue to acquire a sense of how we believe people should handle us. For example, if you only earned encouragement and feelings of affection from your caregivers while you were growing up while you performed well at work, you may step into your adult life thinking that you must perform well in order to be deserving of affection and to gain your partner's treatment and comfort. If your partner turns away or breaks down on you, leaving you feeling unloved, you might easily begin to focus on your own actions, repeat stuff, and check for something you might have "done wrong" for the individual to treat you in this manner. The predominant source for digging up raw places, old wounds, and patterned habits is our interpersonal relationships. These bonds, as well as being alone and abandoned, will leave us feeling connected and cared for. Not unexpectedly, as imago therapy implies that we choose spouses who seem "familiar" to us, our romantic relationships also appear to dig up tired, familiar emotional wounds. It will give us a chance to recover and develop as these old wounds surface in relationships. Imago Intimacy Therapy believes that this is indeed valid. As a well-known author says, "We are born in relationships, we are injured in relationships, and in relationships, we can be restored."

## Picking a Familiar Partner

Imago therapy means that we prefer spouses who remind us of our early caregivers, a blend of their positive qualities and not-so-good qualities. This is an explanation of why we feel acquainted with the person we tend to connect with and why we would be comfortable putting our guard down with them. Since they have features, we are acquainted with, because of what we experienced growing up, we still seem to recognize how to handle certain features. To accept that we might choose a partner with the same not-so-great qualities as an early caregiver might make us mad. However, it makes sense that we prefer to find things easier to handle environments and individuals that appear familiar to us. Compared to someone who is more assertive and deliberately engaging in verbal exchange during periods of tension or anxiety, whether you are used to having the cold shoulder from a caregiver during times of conflict or anxiety, you may find an odd, comfortable warmth in a spouse who does that as well.

## What Makes It Different?

While these ideas are found in multiple forms of dynamic psychotherapy, imago therapy stresses that our early caregiver attachment interactions will specifically impact our spouse's preference as an adult. When we date, almost as though we have met them before or for a long time, we may come across someone that appears all too comfortable and quick to communicate with. What imago therapy implies is that, in our early experiences, these entities seem natural to us because they have been in parallel relational interactions with caregivers previously. We start letting our guard down and develop closer as we feel relaxed and acquainted with another, which makes it easier to establish a romantic relationship. The closer we get overtime, inside our relationship, we can notice old emotional wounds surfacing and question what is happening. Another feature that separates imago therapy from other therapeutic types is that it relies on the utilization of conflict and anxiety, including recovery, including development potential. Imago therapy allows partners to tap into the times of tension and use them for discovery, excitement, and learning, rather than telling others how to actually "battle harder" or find solutions to prevent friction within their relationship. Imago therapy is relational, which means that a psychiatrist may not have a separate function as an expert to provide guidance, but rather that the psychiatrist operates closely with the pair to look at what is happening to them and heal the relationship as a whole. The advisor encourages the pair to be the experts in their dynamics, encouraging couples to benefit from each other in a manner that promotes the dialogue.

**What Can It Help With?**

Imago therapy was designed primarily for relationship awareness and recovery. Many of the conditions for which imago therapy may aid include:

- Challenges in Communicating

- Recurring conflict / disagreement

- Sentiments of disconnection

- Lacking familiarity

- Unfaithfulness/trust issues

To engage in imago interpersonal therapy, you do not even have to be in pain. In fact, partners that are not in crisis will profit greatly from communicating, knowing about these complexities inside the relationship, and having a deeper understanding of themselves and their spouse.

## Who Can Imago Help?

Excellent applicants to benefit from imago therapy will be people in stable relationships with a significant other. From dating and premarital partners to those who have been married for several years, partners at all levels and seasons of their relationship are invited to join. In imago relationship therapy, people may also engage. People who are dating will definitely profit from talking about their relationship habits, partner preferences, and how to meet someone that is a safe partner and a good partner, and how to communicate with them.

### Imago Dialogue

Imago dialogue is a central component of imago relationship counseling. This conversation, mediated by a professional imago therapist, is a formal process that helps partners to develop understanding and improve empathy. The aims of the imago dialogue are:

- Delete from conversation derogatory, hurtful words.

- Creating a safe relational atmosphere for all spouses to communicate freely.

- Enable fair room for all spouses and remove the concept of one spouse getting greater control than the other.

There is a "sender" and a "receiver" throughout this conversation. The sender is the one to freely express thoughts and emotions with their recipient. "During the imago conversation, the" receiver "practices the following three steps:

**Mirroring**

To achieve clarity and comprehension, reiterate what you just heard your partner say. With no opinion, complaint, or reaction, the recipient does this, just merely reinforces what they have heard their partner tell.

**Validation**

The recipient is working to verify aspects of what has been communicated with their partner (the sender), which makes sense to them. When they do this, they let their partner realize that they are "getting it" and are working consciously to grasp it. If there are pieces that are not yet known by the recipient, they may ask the sender to share further.

## Empathy

The recipient discusses with their partner what they believe the other would feel at this stage in the conversation. Talking at this point is a way to let the spouse realize that they are getting a better awareness of their emotional background, helping the spouse to be seen and understood.

## Imago for Individuals

While imago relationship therapy is a type of counseling intended to function best for partners in engaged marriages, to profit from imago therapy, you definitely do not need to be in an active relationship. In reality, many dating individuals may find this form of counseling very effective in analyzing their own past and how it could influence their dating habits and partner choices. You will understand what any of the old wounds or emotional, raw points maybe that harm the relationships by engaging in imago therapy by yourself. It may be helpful to feel a sense of peace around these raw spots and help you step on with more confidence and learn how to be a great, caring partner in your next relationship.

## Common Questions

### How Can I Get Started with Imago Therapy?

Workshops and counseling exercises provide two key avenues to start understanding more about imago counseling and how it will improve the partnership. Several diverse seminars are open. All focused on the imago therapy model. Any of the available workshops are geared especially to:

- Premarital partners

- Couples facing distress

- Couples with kids

- Religious couples

- Same-sex partners

- Single Individuals

Workshops are provided around the world, and workshops in your city or country are likely to be open. In therapy by an imago-trained therapist, the other form of involvement is. Traditionally, classes are provided one hour at a time, but alternative programs are also available, such as intense workshops that last a few hours or retreats that may continue for a few days. In an imago-trained professional, getting face to face interaction helps you and your partner to consciously delve through the complexities of your relationship. During that time, you can use conversation to discuss and learn what happens to your spouse when there is pain or tension in the relationship, facilitated by the therapist. It will improve intimacy and build a sense of link and reconciliation between couples to consciously pursue communication such that the same behaviors and challenges avoid popping up time and time again.

## How Can I Find an Imago Therapist?

Most clinicians who deal with partners are likely to have had some experience in imago marital therapy; and basic knowledge; In your city, you will find services at places like Imago Relationships International, such as qualified and even completely accredited imago relationship therapists. You will scan for a directory of qualified imago therapists from around the globe, looking for your place and level of need for relationships. You will also find sites for a number of available seminars, which are focused on the ideas of imago relationship counseling.

## Are There Times When Imago Therapy Might Not Help?

Like in all forms of marital counseling, there are occasions that the marital will not be a suitable match for imago counseling.

These periods could include conditions such as physical assault, misuse of active drugs, or even addictive habits that may get in the path of a successful outcome of marital therapy. Imago care will only be effective where complications like this are fixed first.

## Sex Therapy with Sensate Focus

Sex therapy is a type of psychotherapy that is intended to support sexual issues to be resolved by people and couples. Female therapy, not hands-on therapy, is conversation therapy. Everyone in the audience is completely dressed in a sex therapy session, although there is no contact. In order to rule out medical causes of sexual issues, sex therapists can suggest obtaining physical examinations. For those who are focusing on sexual issues that may not have a partner to experiment with, they can even promote the usage of sexual surrogates. Sex practitioners, though, should not conduct medical tests on their clients or have intimate experiences with them.

### What Is Sensate Focus?

Sensate concentration was first introduced in the 1960s as a sex therapy method. In order to strengthen their relationship and bond, it includes a set of behavioral activities that partners perform together. An understanding of shared obligation and a desire to perform homework as recommended by the sex worker are two of the most critical factors in the effectiveness of sensory focus. Mutual accountability is important since sexual problems are framed as a concern of the pair rather than a concern of the person that has been described as "the patient." The aspect that distinguishes sensitive emphasis from other therapeutic interventions is organized homework assignments. The trademark of sensate emphasis is that stressful activities are temporarily excluded from the sexual menu of a pair. Then, the doctor prescribes a clear recipe with measures to take to strengthen the sexual lives of the pair with the causes of tension eliminated.

## Foundation of Sensate Focus

There are seven components that serve as the basis for responsive attention. They are:

- Establishing shared actions to address sexual interests and expectations among partners.

- Providing sexual role and sexual behavior details and education.

- Being prepared to modify beliefs towards sex.

- Keeping rid of guilt about sexual success.

- Helping couples develop female contact and sexual strategies.

- Reducing unhealthy relationship habits and sex stereotypes.

- Giving assignments to help partners improve their sexual relationship for the better.

**Sample Sensate Focus Exercise**

Reducing output distress and enhancing communication are two of the key objectives of sensory attention. A standard early homework task may go something like this for a couple where one partner has erectile dysfunction. Therapist would say," For the next week, I want you two to find two evenings that you can spend at least an hour together. One of you can plan the date on the first night, the other on the second. The bedroom with clean sheets, good light, and fun music that you all find soothing will be set up by whoever arranges the date. Every one of you can take a warm shower to relax before your date. You should do so if you'd like to wear panties for this first workout. The individual setting up the date would then assist their partner on the bed to get relaxed. They would then spend half an hour exploring the feeling of touching the body of their partner and loving it. For now, since we want to maintain this encounter low tension, we're going to discourage touching genitals. You're supposed to turn after half an hour. So, the other person can have the chance to do the same kind of experimentation. The purpose of this homework is not to offer a massage to your person. Instead, without any requirements, it is to find pleasure in touching and being touched. That's why it's crucial to connect during this date. Inform your partner whatever you like and also what you don't really like. Let them know what you really want and what you don't like.

**Why Sensate Focus Is Used**

Sensate emphasis has been found to be successful in addressing a variety of different kinds of sexual distress in women and men as a part of sex therapy, including:

- Discomfort during sex

- Ejaculation prematurely

- Erectile dysfunction

- Difficulty in arousal

- Disorders of desire

Sensate Attention is an intervention focused on couples. It may be used by couples, gender roles, and sexual orientations of all various ages. Most of the sample was for married partners. Still, among the same sex and diverse preference partners, several clinicians have embraced it.

## Effectiveness

There is a lot of research exploring the usage of sensible emphasis, alone or in accordance with other approaches, to increase the sexual pleasure of partners. For those struggling with sexual problems as a consequence of medical disorders, such as breast cancer, the procedure has often been utilized as a part of sex therapy. Sex therapists and other clinicians dealing in sexual illness are well-accepted by Sensate emphasis. That's especially true when it is used in tandem with good sexual behavior and work education. A really safe strategy is Sensate concentration, and most people find it easy to obey. That is, in large part, because the sensible emphasis is deliberately intended to minimize performance anxiety and tension about sexual activity as a gradual and gentle process. Many sex therapists report that responsive attention is a clear and successful way for partners, same-sex and opposite-sex alike, to improve trust and attachment. However, utilizing sensate emphasis, not all partners or person therapists are relaxed.

## How to Find a Sex Therapist

There are a couple of options to locate a sex therapist. Searching for the insurance company's directory of suppliers is also the

most affordable. Seek a specialist in mental health that special-
izes in sex counseling. Therapist listings and sex therapists
should even be checked and cross-referenced with the insurance list. Finally, a directory of sex therapists is compiled by the
American Association of Sexuality Educators, Psychologists,
and Therapists on their AASECT.org page. To receive both
preparation and clinical supervision in sexual health and counseling strategies, AASECT licensed sex therapists are required.
That said, there is also time-limited sex counseling. It is commonly anticipated that pure sex counseling would last for
longer than 10-12 sessions. The number of appointments you
require, though, can differ based on the concerns you are trying
to resolve and whether you already have a general counseling
sex therapist.

## Partner Yoga

Per year, over 36 million Americans say "om" to yoga, and with
good cause. The relaxing, toning exercise can be a perfect relief
from everyday life's stressors, whilst improving your endur-
ance and power to boot. And the advantages go well beyond
just chiseled bodies and solid glutes. Studies suggest that the
practice will aid in anything from insomnia care to the avoidance of diseases such as diabetes. Yoga is historically a disci-
pline for the practitioner. At the present time, it's a chance to

stretch, relax, and reflect on your mat. However, it may bring

its own special advantages to perform yoga with another individual jointly, whether it be a partner, spouse, or significant other. Identified as yoga for partners or pair yoga, this exercise helps two persons, by supported poses, to connect to each other. From extending your faith levels to improving your bond, the yoga of couples will have a positive influence on your relationship that goes way beyond the physical.

The yoga experience of a couple will act as a sort of mini "retreat" or "workshop" to improve a relationship. Instead of only heading to a class and exercising next to each other, pair yoga allows partners to actually pay attention to each other at the moment and strive for mutual interests together. For all partners, the exercise is mutually advantageous, and research suggests that yoga for couples has far-reaching effects, including lowered anxiety to happier sex life. Read on to hear more about the special physical, social, and behavioral effects of performing yoga for couples.

### Is Partner Yoga, the New Couples Therapy?

Learn how partner yoga will change your relationship with your partner positively. No pair is resistant to coupledom's difficulties. Although counseling may be the standard go-to resource for encouragement, certain counselors with yoga experiences take sessions on the pad, empowering individuals through partner yoga therapy to improve their bond through asana, pranayama, and meditation.

### Using Our Bodies to Bypass Our Brains

Couples can synchronize their breathing during sessions, assist each other with asana, or bind their bodies together to establish one pose. With the cooperative nature of the counseling, couples are expected to focus on each other, which makes coordination important and creates trust in the method. Where conventional talk therapy focuses on conversational insight advantage, somatic-based approaches place the mind and body in concert to resolve well-being via guided action, perception of bodily feelings, and different poses in the case of partner yoga.

**It's a scientifically-backed approach.**

Research suggests that this form of body-oriented psychotherapy may alleviate tension, decrease symptoms of depression, and reduce anxiety, whereas yoga practice may increase sexual intimacy, increase relationship fulfillment, and foster compassion. Couples also justify their actions while disputes get serious, which lets the point of view-taking and compassion goes sideways. Partner yoga brings us into the body, which helps the nervous system calm down. Finger-pointing softens, and partners will witness the interactions of each other with more compassion in a more comfortable state. Lying about topics can hold us in our minds, but spouse yoga teaches partners how in actual life, their contact habits turn out. For e.g., when couples are embroiled in reactivity and emotion, the teacher can ask one individual to rest in Balasana (Child's Pose) with the hand of their partner softly resting on their sacral region. In order to improve the position, the person on the mat must then convey what they need from their partner and then give more input on the improvement. After both people have a turn keeping the pose and giving encouragement, the teacher draws focus to some strong emotions that could have been present only moments before the exercise: "When did the tension surrounding your rage change? What have you noticed?

### Using Simple Moves to Deepen Closeness and Ease Conflict

Social sciences term this an experiment inexperience. Including mindfulness, mediation, and art therapy, yoga therapy for partners depends on body activity to further reveal interest and insight into human actions. And to benefit from this type of yoga, couples do not need to be seasoned yogis. Minimal physical prowess is needed for this form of yoga care. What is more necessary, for both yourself and your partner, is a desire to turn up. Breath and contact sharing help us to be more involved in our minds through our bodies and feelings rather than in a reactive position. In this way, yoga allows partners to gain an understanding of their feelings, utilizing nonverbal actions in their gestures as markers to connect physical activity to mental activity. Who, in fact, offers insight into the dynamic of the pair?

## Strengthen your Body and your Relationship: Benefits of Couple's Yoga

### Increased Relationship Satisfaction

Only the act of checking out your partner's yoga class for a couple will make you become happier with your relationship. Studies have found that partners who participate together in challenging new experiences will feel an improvement in both the strength of the relationship and emotional appeal. Furthermore, the intimacy and mutual posing in the yoga of couples will help to refresh and restore a bond. It helps couples to have fun whilst calming down, enjoying quality time, and having a positive relationship while gaining different skills together. Yoga often generates consciousness; which research has related to better experiences. The research reported that in a positive correlation between enhanced mindfulness, described as accessible attention to and mindfulness of the present moment, and greater satisfaction with relationships. It will reinvigorate the bond by being in the present moment while concentrating on breathing and posing with the partner, helping you both feel satisfied with your relationship.

**Improved Intimacy and Sex life**

Yoga for partners will also serve to improve both arousal and sexual pleasure. Partner yoga can support partners who are dealing with sexual dysfunction, a study showed. It's important to remember that yoga for partners is not erotic in nature. It's a form of yoga that uses two individuals to synchronize their breathing, postures, and gestures. However, since it needs new forms of faith, contact, and interaction, this will improve intimacy. Due to improved contact by touch and action, one explanation of why yoga will enhance your sex life is. Relationship tension may arise from partners feeling out of touch, detached, or disconnected. The act of traveling together will make couples feel more in harmony in pair yoga. Studies have demonstrated that yoga exercise will increase sex drive and, in reality, some pair therapists are already introducing partner yoga into their therapy sessions to help partners strengthen their sex lives and develop a better relationship.

## Increased Communication and Trust

You must support and depend on your spouse throughout (both physically and metaphorically), as well as continuously interacting verbally and nonverbally, in order to construct the poses in a couple's yoga session. This needs trust, assistance, and, most significantly, weakness. Physical touch may be a language of its own, a means of conveying, without using words, a sensation of nurturing, and sharing deep feelings. The capacity to convey to another person that they are heard, respected, cared about, cherished, acknowledged, supported, deserving, and protected is aware and consensual human contact. In comparison, synchronized nonverbal motion, such as that seen in the couple's yoga's rhythmic breathing and posing, will make couples feel "more affectively attuned to each other." According to the report, copying your partner's movements, often called mimicry, will help to improve intimacy and bonding. This will help to strengthen coordination since couples need to focus on each other to remain calm and solid in poses. The flowing postures, the push and pull, and the dependency on someone else create a relationship, so the moment and the gestures must be completely engaged by the participants.

## Reduced anxiety and stress

Although most yoga activities help to relieve tension and reduce anxiety, couple yoga gives a special benefit due to the strength of the contact of your significant other. An analysis showed that married couples keeping hands feel relief from intense tension instantly. The spousal hand-keeping gave a greater neural reaction than keeping a stranger's hand. Thus, by helping to reduce the neuronal reaction to stress, just touching your partner will decrease anxiety. In addition, certain poses are intended to help loosen up some parts of the body, such as backbends and Camel Pose. This will provide space for fresh energies and offer relief from tension, tension, and discomfort, both physical and mental. If you perform yoga to alleviate stress, develop power and flexibility, work on concentration, or a mixture of both, couple's yoga has the added advantage of improving your relationship. And that is everything about which we would tell Namaste.

### Back-to-back breathing

- Sit in a relaxed posture facing your partner, with your backs leaning against each other and your feet in a cross-legged position.

- Stand up straight, shoulders straight and away from your face, arms relaxed, and start breathing alternately.

- When your partner profoundly inhales, you profoundly exhale, and so on.

- Repeat and repeat three times for ten breaths.

This breathing pose can help improve your partner's mindfulness, calming, and communication.

## Couples Yoga Poses to Strengthen Your Relationship with Your Partner

These partner poses will help you build a closer bond with your other half. When you are married or have a long-term relationship, it may be challenging to find new ways to communicate with your other half and expand your relationship. Yeah, there can be surprise tickets for concerts and night out dinners that involve you saying, "I am in love with you,". But you can't do it over a plate of lasagna to get to the deeper stuff, including creating confidence and intimacy. A central feature of partner yoga workouts is coordination. A perfect way to build a basis for transparency, trust, and empathy of one another is to connect properly during your partner's work. Being transparent and truthful about each other and maintaining things light-hearted and enjoyable is the most crucial thing about partner yoga. Let us show you some yoga poses for couples, which are intended to help improve the relationship between partners. Take your time by relying on your breathing. Making sure both of you speak about each posture with each other to ensure that you are in the best place and that you experience the stretch properly.

**Seated centering**

This is an ideal way to begin every practice in yoga. It helps you to interact and brace your mind and body for the work you're about to initiate with your spiritual and physical setting. In a good yoga routine, attitude and reflection are essential elements. Put your hands on your partner's knees, sit cross-legged, facing your partner.

If it is difficult to sit cross-legged, sit up for further space on folded towel/cushion. Look at the eyes of your partner and take a few seconds to actually see your partner. In and out, take ten deep breaths and make a deeper communication without words. In our everyday lives, we may get so distracted that we neglect the value of really looking at each other in the moment as we are.

**Seated cat-cow**

These are two poses in yoga that are commonly combined together. For the hip, the core, and muscles in the back, it is a perfect stretch. This pose often assist with lung and chest expansion, so make sure to concentrate on your breathe while practicing this position. Be seated and aim for the forearms of one another. When you release and expand your shoulders back and down, keep a strong grasp. Raise your chest and shoulders towards the ceiling as you inhale, causing your upper-middle back to get a subtle arch. Draw your chin towards your chest when you exhale, rounding out your upper-middle back while pushing your shoulder blades apart. For a few rounds of air, follow the same action, and when you warm-up your back, when you inhale, you should raise your eyes towards the roof and encourage your head to sink to your belly-button when you exhale. Practice ten to twelve rounds to gain trust when finding a sensation of lightness in your upper back and chest using your partner's encouragement.

**Back-to-back chair pose**

For beginning yogis, this one is a perfect pose since they can use each other's help. A perfect way to reinforce the muscles of feet and the thighs while the stability of the ankle is the chair position. Rely on each other for intensifying the stretch with more experienced Yogis. Keep your arms relaxed and, on your backs, stand back-to-back. When you shift your feet apart on hip-width, then move slowly away from back of your partner. Bend both your knees and then lower yourself gently, assuming you were seated on a throne. When the knees have hit an angle of ninety-degree, stop and take eight to ten steady breaths. Make sure that your head is raised like a crown and that the length of the neck is preserved, pushing equally with both feet. Push onto each other and straighten your legs, and transition back. This practice helps create confidence, especially when you are using the support from your partner during the shift from standing to the chair.

**Seated forward backbend**

For the legs and back, the forward bent and backwards bend are supposed to be one of the extreme stretches. This pose is a little difficult, especially if first partner is more agile than the second. So, make sure you connect and take things easy, or it can affect you both.

Lay down for your partner, with both of you taking support of each other's back. Help one partner stretching legs and start folding over. The second partner lowers the legs and puts the two feet down on the concrete, starting to lean sideways on the other person's strength. Keep for a deep breath of five or six, then come upright and turn places. Based on their reviews, you should add further pressure after checking in with your partner on this one. When you lean towards each other, this pose culti-vates physical interaction.

# Chapter 4: Improving Intimacy, Appreciation and being Mindful

"Couples counseling" and "couples therapy" are usually considered as the same thing. On a scientific basis, there is little distinction between them. The other way in which whether the session is called matters is a valid one; in certain places, you may get a separate "therapy" qualification or license to practice that is tougher to receive than the "counseling" qualification or license to practice. This form of relationship with a trained therapist presents partners with an ability to work on their most complicated or socially demanding concerns, whether you name it partners therapy or couples counseling. These topics may vary from basic difficulties of understanding or serious disputes to difficulties of drug misuse and psychiatric conditions. Although counseling for partners may be a wonderful way to bond with your spouse or mend the gaps between you, without having a therapist, there are also ways to ensure sure you maintain the flame is going and the relationship safe. There are several tools out there that rely on couples counseling ideas or tests. It is never too late to start adding a little more time into your relationship (or too early). Choose one or two of the practices and tasks listed below to practice with your spouse if you would like to strengthen your relationship. If there is a fully efficient tool out there for stable, safe marriages, certainly somebody might have picked it up and marketed it by now, right? We will have to do with what we have before we can discover the 100 percent performance promised blueprint for a perfect relationship, building our communication skills, connecting efficiently, participating in events the strengthen our bond, and utilizing couples counseling to resolve some of the major issues. In order to develop a stable relationship and ward off divorce or breakup, there is no "best" behavior for partners because each partner would have their own best practice. For certain couples, it might be joining together in a common sport, such as riding a horse, playing a classic series, or playing guitar together. For some, when gazing up at the night sky, over coffee in the morning, or lying in bed at night, it could be the prolonged chats they sometimes have.

## Exercises to Improve Intimacy

There are also faster and easy ways to understand more about your relationship and strengthen your bond, often helped by couples' counseling and therapists. How complicated relationships can be is plain to see. Keep reading to learn more about all of these wonderful approaches to create a great partnership and sustain it.

## Soul Gazing

This is an intense activity that can help you communicate at a deeper level with your partner. It will have a tremendous influence on the sense of connectedness, so it is not for the faint of heart. Face your partner in a sitting posture in order to attempt this workout. Step so close to each other that the legs are almost touching, and stare into the eyes of each other. Keep for three to five minutes of eye touch. Don't worry, and it's not a race. You might blink. Furthermore, refrain from communicating. Simply gaze at each other's eyes, even though at first, it is uncomfortable. Choose music that is fun to all of you or significant in terms of your partnership and keeps eye contact before the music stops, if the pause is awkward. Even mainstream media has gained an appreciation of this exercise's strength.

## Extended Cuddle Time

This is just an easy and enjoyable workout, as it seems. The directions are clear to cuddle more often. With a mobile phone, tablet, or book at bedtime, it is easy to get frustrated, but cuddling is really a lot healthier way to finish the day. When we cuddle with our partner, the chemicals that are emitted boost our attitude, strengthen our bond, and may even make us sleep better. This exercise is supposed to be practiced just before bed, but if bedtime does not fit for you, you should carve out every time of the day to cuddle. The main thing is to have some one-on-one attention, display physical affection, and develop your partner's intimacy.

If you have difficulty locating or sticking to a daily cuddle session, Relationship Counselor advises cuddling to a music playlist. When enjoying a movie, you can even sneak in some cuddle time or first thing every morning when you both get up the point is to fit it in, so it fits better for you.

**The Breath-Forehead Connection Exercise**

This activity is an outstanding way of taking your mind off what's happening around you while reflecting on your partner. To start with, either lie down with your partner on your side or sit up straight with your spouse. Face one another and bring the foreheads together softly. Make sure the chins are turned down, so you don't bump your noses and remain for a few breaths in this place. In harmony with your partner, breathe at least seven long, deep breaths. At first, it might be challenging, but before long, you'll get the hang of it. If the activity is liked by you and your partner, feel free to extend it; take up to 20 breaths together, or 30 or only breathe together over a certain period of time. There are no negatives to feel linked with the partner, so go for it. You and your partner would be placed into an intimate, linked room by this near breathing exercise. Whenever you notice the urge to calm down and refocus on one another, practice it.

## Exercises to Improve Communication

### Uninterrupted Listening

Uninterrupted Listening is another easy but effective practice, and it's exactly what it looks like. We just deserve to be heard, understood, and cared about, and both you and your spouse will feel this way through this exercise. For this exercise, set a timer (usually three to five minutes would do the trick) and let your partner chat. Jobs, education, you, baby, partners or relatives, stress; it's all fair game. They should chat about whatever is on their mind. Your job when they are talking is to do one thing and just one thing: to listen. When the timer goes off, do not talk at all. Listen to your partner plainly and take it all up. Although you do not talk at this moment, by body language, facial expressions, or meaningful looks, you are free to offer your partner non-verbal support or empathy. Switch positions as the timer go off and try the exercise again. You can notice one partner be even more chatty than the other, which is perfectly natural.

## The Miracle Question

For partners, this activity is a perfect opportunity to discuss the kind of future they want to create, separately, and as a couple. Often, we all struggle, but sometimes the challenge is worse because we really don't realize what our ambitions are simply asking the "Miracle Question" to help you understand your objectives. This query encourages all participants to test their own dreams and desires and think about the dreams and desires of their spouse. In order to be content with the relationship, it will allow a person to consider what they really need and their essential other needs. The Miracle Issue is typically phrased this way by therapists: "Suppose there was a miracle tonight when you were asleep. Tomorrow, when you wake up, what are some of the stuff you'd find that will convince you that your life instantly got better? While either spouse can offer an answer that in their waking life is an impossibility, their response may also be beneficial. When practiced in the sense of the counseling of families, the psychiatrist will delve further into the unrealistic miracle of the clients with this question: "Why can that make a difference?" This dialogue allows partners to imagine an optimistic future in which they solve or minimize their challenges. If you are involved in this activity without a therapist's help, if it is impractical or impossible, don't want and dig too deep into the response. Use this conversation, however, as a chance to discover something different about your spouse and work together for your future.

**Weekly Meeting**

This activity would be a perfect opportunity to interact if you and your partner are living lives jam-packed with tasks, events, and responsibilities. This activity presents an opportunity for you and your partner to connect as adults (no children allowed) and without disruptions (no mobile, tablets, or laptops allowed). Plan an unchangeable block of time (30 minutes is a reasonable default) once a week to speak to you and your wife about how you all feel, your relationship as a couple, any unanswered disagreements or issues, or any unmet needs.

You will begin the exercise with questions such as:

- How do you feel now about us?

- Is there something that you would like to chat about from this last week that you find incomplete?

- How in the coming days can I help you feel more

85

adored?

In a safe and fruitful conversation concerning yourself and your relationship, the responses to these questions should direct you and your spouse. To stay on top of any concerns, make sure to do something periodically to guarantee that items are not swept under the rug or placed on the downside for too long.

**Five Things to Exercise**

This exercise can be carried out wherever the two of you are together, as another fast and simple exercise. It is just your vocabulary and your creativity that you require. For each time that you do this exercise, come up with a theme. Anything like "what I'm thankful for," "what I admire about you," or "what this month I'd like to do with you,"; and inside this theme, mention five items each. You may make one partner go first and mention all five items, or you and your partner may say one of your five things at a time alternately. You plan to do so, however, to be creative and not be scared to get stupid with your partner. For starters, you may ask your wife, "What are the five things you love that I have been doing for you lately?" Their replies may be like, "Take out the garbage, make a reservation for dinner, get my car in detail, cuddle with me, and watch my favorite movie with me." "When they've done their list, come up with your own answer to the issue, such as," Replace the water pump, pull the weeds, stitch the button back on my dress, remind me how much you love me, and kiss me every night. You should chat about your things after you have each done discussing your list, show each other gratitude, pose follow-up questions, or come up with more things together. This activity is a partnership and enjoyable way to communicate, discover something new or reminisce about good mutual experiences with your spouse.

## Exercises to Improve Appreciation for Each Other

### Appreciative Inquiry of Relationships

In a romantic relationship, if you are struggling, this is another good worksheet to try. It may also help to look further into the positive aspects while a pair is having difficulty, rather than the difficulties they are facing. Appreciative Inquiry (AI) explores, by constructive questions and polite inquiry, what brings vitality to a partnership. This technique will be used by a pair to open up their history and look at their accomplishments, potential, beliefs, and talents as a pair. The pair would be encouraged to note that they are a team with similar interests, similar desires, and common characteristics.

### Identify an Important Relationship

For deep dive into family dynamics and relationships as well as for intimate partners, this AI worksheet is perfect. In your comments, strive to be as specific as practicable, pointing at the present condition of the relationship and your thoughts against your partner and stuff between you.

**Discover**

This is split down into two steps. First, a celebration exercise in which you can note why you would like to enjoy a common event. What was this event or moment worth celebrating? What is it that has rendered it so beneficial? What attributes have you taken to this moment? How about a buddy of yours? Next, a maintenance task in which you can focus both on your own and your partner's most positive commitment to continuing the relationship. What do you both offer in a safe way to keep the relationship developing? What does that contribution actually work for?

**Dream**

Your next move would be to mutually imagine your dream future. Do you find the time and shut your eyes and see as you want it to be? For your future together, what are your goals and dreams? What terms might be used by your wife to identify you that will make you feel proud? To explain your dreams, use the space beneath this phase.

**Design**

The next move for your idea freshly written down is to design tangible measures that you will take toward the perfect future yourself. Think of your personal qualities and how they will help you reach your dream relationship, and then adapt the same technique to the actions you and your partner will take together.

**Destiny**

To set forth your intentions so that you can stick to them, use this substantial final space. Know the reason behind each intention you are going to stick to, and this can further inspire you where there are some possible barriers. By connecting them back to your own beliefs, as well as why the relationship itself is so important in your mind, strive to render these "whys" as meaningful as possible. Filling in these blanks can allow a person in their relationship to consider the good stuff and contribute to substantive progress that constructively draws on their talents.

## Apologizing Effectively

In the conventional context, this is not necessarily a worksheet, but it contains useful details on how to properly apologize when any person has upset their spouse or undermined the faith in a relationship. It is too great a resource not to share for this purpose. The four measures to apologize effectively are set out as follows:

- **Acknowledging**

You may initiate an efficient apology by accepting responsibility and admitting the related crime, whether you have harmed your spouse intentionally or unwillingly. Prove that through "I" ("I screwed up ..." or "I am at fault ..."), you acknowledge your obligation. Recognition of who was affected, as well as the essence of the transgression itself.

- **Give an explanation for the offense.**

Explain both that you never wanted to harm the other person and that in the future, it won't happen. This worksheet also gives suggestions about how to differentiate between excuses for effective, substantive apologies and explanations.

- **Express your remorse**

Naturally, among other emotions, we feel guilt and remorse when we harm another person. It will help your wife appreciate your recognition of the error by voicing the feelings that you experience, such as humiliation, remorse, embarrassment, and so on. "I feel very guilty about what happened, for instance. I have been embarrassed for days over how I let you down.'

- **Make amends**

Follow up with your verbal recognition and apologies with behavior directed at repairing the harm incurred. Speak to your spouse and see if they may consider your partnership to be a reasonable reparation for the hurt.

## Naikan Reflection

While this Naikan Meditation worksheet is something for each spouse to focus on individually, for couples who would like to develop and sustain a stable relationship, it is also a fantastic guide. The exercise prompts the reader to try not to take for granted his or her partner. It helps to foster feelings of love and respect while helping each spouse to become more mindful of their partner's spiritual status. Following these guidelines will help spouses display respect for their partners and bring more gratitude into their relationship. Naikan Meditation is a Japanese self-reflection approach that has three questions in therapy; there is room for you to document the responses on this sheet. Looking back about the past 24 hours, and especially with your partner in mind, focus on the following:

## What have I received?

What encouragement, treatment, and attention did your

partner owe you? To raise your self-confidence, did they tell kind things? Driving you to work, or afterward, pick you up? Is your lunch packed? Feel free to mention all that you have got over the past 24 hours from your wife.

**What have I given?**

Have you been telling them how school was? Have you contacted them only to say hello? Will they pick up their clothing or scrape their shoulders?

**What troubles or difficulties have I caused?**

Think of all the aspects in which you might have created a problem or damage your partner during the day. Maybe you screamed at them out of impatience, or did you point out what they have forgotten? Did you doubt either of their work or did you forget to do anything they depend on? The aim of this exercise is to build the partner's inherent love to serve and care about them. Through increasing the recognition of how much you already receive and giving in return per day, Naikan Meditation is a helpful activity to increase the sense of gratitude and its benefits. A summary of couples counseling or couples therapy has been discussed so far, and some entertaining, insightful, and supportive activities have been developed to make the best of a romantic relationship. We hope you've found a few different insights on how you should connect with your partner, but we also hope you've got the underlying message; there's no ideal relationship, but when it's tough, there are perfect ways to stand up for each other. No relationship is without an occasional challenge, and any deliberate work on the part of each spouse will help even the strongest. There is much to think about your partner and more fresh and exciting stuff to do together, whether you are in a new relationship or moving to your 50th anniversary.

## Mindfulness Therapy for Couples

There we've always been. Your partner does something that rubs you or vice versa in the wrong direction, and you initiate an argument. Blood pressure increases, stomach-churning, fingers clenching as you all become gradually revved up; you lose the ability to think clearly or consider each other's point of view, a warning of the cortisol levels. The tension hormone, the body-brain, is running heavy. Either or all of you bursts before you realize it and says something jaw-droppingly mean, something you will never take back. What should have been a small argument has developed into a huge dispute, and you both walk away from your partnership, feeling wounded, frustrated, and unhappy. And the discontent could contribute to more tensions in the future and more heated tensions.

## How can Couples Avoid this Vicious Cycle?

Some optimism is provided by a new report. Eighty-eight romantic partners were studied by researchers when they explored a disagreement in their relationship. The researchers took saliva samples from both partners before and after the dispute to assess their cortisol levels, an indicator of how de-

pressed they were feeling. The researchers immediately afterward have questioned each partner what they were feeling during the dispute.

In fact, they needed to learn how well the partners discussed their disagreement of empathy, moment-by-moment, "non-judgmental" empathy of emotions, perceptions, and sensations; a willingness to recognize and acknowledge what we are witnessing in the now, without evaluating certain thoughts or feelings as "right" or "wrong." The outcomes revealed that the cortisol levels of spouses usually spiked during their confrontation, an indication of high tension. But those with better knowledge tended to heal more easily. Since the dispute concluded, their cortisol levels were faster to return to usual, meaning they were holding their cool. For both males and females, it was real. Mindfulness lets couples control their own reactions and embrace each other more fully. When it happens, it leads to less harmful fallout from confrontation. Why do these advantages bring mindfulness? Further research found that during the confrontation, mindfulness helped romantic couples not take it too negatively, control their emotional responses more easily, and empathize more intensely with their partner. The researchers believe that when mindfulness encourages people to be more active in positive disputes, it also encourages them to disengage from disputes that become disruptive more easily. In the middle of the dispute, how could these effects be extended to couples?

## Pay attention

The first is called "attentional sensitivity," which implies how much each participant during the confrontation will deliberately tap into his or her thoughts, feelings, and body experiences without becoming too caught up in certain thoughts or emotions, rather than watching them from a safe distance. "In the report, persons who received high recognition during the dispute usually identified with claims such as," I was conscious of my thoughts and emotions without over-identifying. One of the most successful interventions that stimulate focus is clearly to inquire, "What are you noticing now?" It has been discovered that by merely transferring the mind of the partner to their own body stimuli, clenched jaws or fists, tightness in their neck or chest, churning in the stomach, there may be enormous benefits. And by marking their emotions when they emerge and escalate: the most prominent ones are the rage, depression, fear, and shame. Noting their thinking and action habits allows us to recognize them for what they are: normal and automatic, well-grooved through the neuronal circuitry of the brain. And these habits should not need to own or characterize us, like any habit; they are things that we can alter.

**Be accepting**

Then each partner may be driven by "attitudinal mindfulness": being accessible, engaged, curious, embracing their own experience ("mindful curiosity" is often alluded to by the researchers). "In the report, associates who ranked high in mindful inquiry usually identified with statements such as" I was interested in each of the thoughts and feelings I had. Noticing and naming feelings, impulses, ideas, and acknowledging them as part of being individual, makes a couple not so individually carry on their own experience. In a relationship that they depend on for protection and warmth, they should understand that if they feel endangered, it is not shocking that they will show evolutionarily hardwired and habitually conditioned reactions, such as hostility (negativity) or withdrawal (abandonment).

**Engage with your partner**

Then change the subject of your mind softly to the perspective of your partner. Curious, they may be. They can listen to the worries, fears, and desires of their spouse. Rather than defending against or threatening their partner, they should re-engage. After having requested each other to be curious about your own perspective first, it is crucial that you do this, allowing each of you the opportunity to control your own emotional condition to the extent that in the next two minutes, you don't feel like you have to win some debate. Then, as you reflect on the perspective of your partner, you would be more able to retain the open-mindedness. There's the magic time when partners (rather than make the argument against each other) turn and look at each other. There could be a light contact to the arm or Kleenex's moving, or just a softer look and a cautious smile. In their relationship of interest and transparency, pay attention also to this moment. And to recognize that you have been willing to move towards a more accessible, responsive role from a habitual pattern of confrontation. One critical nuance of this strategy is that the researchers observed that in various circumstances, the two separate types of mindfulness, attentional, and attitudinal, appeared to have advantages. If a spouse participated in negative behaviors during the conflict; for example, by expressing indignation or annoyance, by insulting their spouse, or by suggesting that their partner alter their feelings or acts in any way; if he or she scored high in conscientious curiosity, the partner who was exposed to such negative behaviors benefited from tension quicker, regardless of his or her degree of attainment. Perhaps it was because addressing the dispute with honesty and curiosity "can enable partners to stay constructively linked during a dispute with their own and the emotions of their partners." However, when one partner actually retreated and failed to work through the dispute, those who ranked high in attentional mindfulness (though this was only valid for men) were the most robust other couples. "It might be that compensating for spouse disengagement by distancing oneself from what is occurring helps," they write, "to retain a sense of equilibrium and/or to empathize with the viewpoint of the spouse." When one's partner is especially verbally abusive, the researchers acknowledged the drawbacks of mindfulness; in and of themselves, mindfulness certainly will not undo the effect of highly harmful. This recent study shows how clinicians, or even partners themselves, may use mindfulness to calm off a confrontation and re-warm their relationship under less threatening circumstances. All partners can heal more quickly as each spouse can respond with greater equanimity. They regain a greater acceptance of the other, restore their ability to take stock and connect, and thereby more easily settle conflicts until they spin out of reach.

## Principles of Effective Couples Therapy

By upholding these fundamental rules, make your relationship work. You may believe like there is no way out of your fraught relationship if you're part of a couple in distress. Myths regarding the poor effectiveness rates in medication and therapy for partners just cause the case to seem greater than it is. Couples frequently refuse to pursue intervention until very late in the game, and either or both might have agreed to call it quits by then. It is also true that becoming a successful therapist for a pair needs different abilities than those needed to be a productive person therapist. Pair counseling may have demonstrably beneficial outcomes when correctly performed. A major analysis of over 40 years of literature on couples' counseling has recently been conducted by UCLA psychologists, in which they synthesized the techniques of the most effective intervention strategies. They also boiled down this vast amount of study to prove that spouses will profit from therapy that meets five underlying concepts through major theoretical orientations within the sector. While one therapist can adhere to a therapeutic approach and another to an emotional approach, both practitioners will achieve meaningful and meaningful change as long as each utilizes common techniques to assist their clients. Evidence-based methods, whether for people or for families, are essential to recognizing successful treatment. This suggests that the medication you are seeking has been checked, ideally in randomized controlled experiments, against similar treatments. Psychologists that have evidence-based counseling should not conform to a single theoretical orientation only because in graduate school, they studied it. Instead, they change their methodology, both therapeutic and science, to ensure that they pursue the latest data. Unfortunately, the opinion of the population is strengthened by media and films that therapists struggle so greatly from their own individual shortcomings thàt they cannot deliver adequate treatment. It can be draining to be a couple's therapist. There is less time to step back, think, and have an answer to a client's comments relative to individual counseling. The session will turn into a screaming match if you sit back for too long. It needs unique expertise to be the therapist of a couple, so that is what the preparation is for. Individuals who undergo years of intensive coursework and monitoring to go through marital and family counseling or therapy go through an exhausting training and certification and licensing phase and have to undergo instruction during their professions and hear about the latest advances throughout the sector. Self-selection is ultimately involved with who chooses to become a marriage counselor and, even more so, who continues in the field. The odds are high that the psychiatrist of the pair you meet is someone who offers this therapy, and he or she is dedicated to having partners achieve meaningful life adjustments. Let's shift now to these five core concepts of good counseling for couples, which are:

**Changes the views of the relationship**

The psychiatrist tries to make both parties view the interaction in a more realistic light during the therapy phase. In a phase involving each partner, they learn to avoid the "blame game" and only focus on what applies to them. They will also profit from seeing a certain sense in which their relationship takes place. Couples that are financially suffering, for example, maybe under various kinds of situational pressures than others who do not. Therapists initiate this phase by gathering by observing how they communicate "information" on the relationship between the partners. Therapists then devise "hypotheses" about what variables can relate to the way the partners communicate. Based on the basic theoretical orientation of the psychiatrist, how they communicate this detail with the pair differs. For a range of methods, from behavioral to insight-oriented, there is scientific evidence. Different clinicians will use different techniques, but the pair will continue to view each other and their experiences in more adaptive ways as long as they work on improving the direction the relationship is understood.

## Modifies dysfunctional behavior

Successful therapists for families aim to improve the way the participants actually conduct each other. This suggests that clinicians will make sure that their clients do not participate in behaviors that may inflict physical, social, or economic damage in addition to helping them enhance their relationships. Therapists would perform a thorough review to decide whether their clients are, in reality, at risk in order to achieve this. For example, if appropriate, the psychiatrist may suggest that one spouse be sent to a domestic violence shelter, specialist therapy for substance dependence, or control of rage. It is also likely that the pair can profit from "time-out" protocols to avoid the worsening of tension if the danger is not too serious.

## Decreases emotional avoidance

Couples that stop sharing their private emotions are at higher risk of being disconnected and falling apart emotionally. The therapists of successful relationships help their clients get out of the feelings and emotions they are fearful of communicating with the other individual. Therapy for spouses focused on attachment helps the participants to be less fearful of voicing their desires for closeness. According to this opinion, certain couples who have struggled to build "stable" childhood relational bonds have unmet needs in their adult relationships that they bring over. They are reluctant to show their partners how badly they need them, and they are frightened that they may be rejected by their partners. Behaviorally oriented practitioners believe that people might be unable to communicate their true feelings because they have not experienced "reinforcement" in the past. Either way, all theoretical methods recommend having their participants reveal their thoughts and feelings in a fashion that can hopefully get them together.

## Improves communication

One of its "three C's" of intimacy is to be able to connect. Both efficient treatments for families concentrate on having spouses interact more efficiently. This dialogue should not be coercive, centered on concepts # 2 and # 3, nor should couples mock one another as they share their true feelings. Couples can also need "coaching" to learn how to talk in more respectful and understanding ways to each other. The psychiatrist can also give didactic instruction to the pair and provide them with the framework for understanding what forms of contact are successful and what forms would likely create further tension. They will learn how, for instance, to listen more consciously and empathically. However, it is important that clinicians refer to the tests they completed early in therapy, specifically how to reach this phase. A different strategy could be needed for couples with a long history of reciprocal criticism than for those who want to prevent confrontation at all costs.

## Promotes strengths

Successful therapists for couples draw out the positives in the relationship and develop resilience, particularly when counseling draws to a close. Since too much of the treatment for partners requires concentrating on trouble issues, it is possible to lose sight of the other issues in which the functioning in partners is successful. The goal of fostering intensity is to encourage the pair to gain more satisfaction from their relationship. The behaviorally focused therapist can "prescribe" anything that one partner performs that pleases the other. Instead, therapists with other orientations who concentrate more on feelings can help the pair construct a more hopeful "plot" or narrative regarding their relationship. The advisor can, in this situation, stop attempting to place his or her own twist on what defines strength and let the pair determine this. We will see, then, that although their condition looks hopeless, individuals in difficult marriages should not give up in despair. In the same way, persons fearful of joining long-term marriages will be motivated after discovering that it is possible to repair strained relationships. These five concepts of efficient counseling, thinking at the flip side, propose directions in which partners can establish and sustain near healthy relationships. In order to further minimize unhealthy habits, take an objective look at the partnership, feel like you should express your feelings, interact efficiently, and reinforce what works. Most notably, you would be offering yours the best chances for success by recognizing that each relationship has its own difficulties and strengths.

# Chapter 5: Communication, Interdependence and Infidelity

Most of us, especially in our intimate relationships, respect interactions with others. In reality, we are wired for communication, and it helps us to establish a bond with our spouse and to establish intimacy. Long-term relationship performance relies heavily on the nature of our intimate relationship with each other. When we think of our perfect relationships, we always think of our most significant partner as a beautiful, close, lifetime partnership. How can we create a partnership of that kind? That cozy, secure, long-lasting bond with someone we know has our long-lasting back? A relationship that allows us the freedom to be ourselves, that facilitates our progress and enables us to be flexible with each other? Understanding the distinction between interdependence and codependence is one of the main factors. It may be difficult to negotiate the depths of marital therapy when one spouse has become unfaithful. Historically speaking, holding a mindset of unconditional good regard towards both the betrayer and the deceived is the trickiest aspect. Very predictably, the psychiatrist tends towards showing the strongest sympathy towards the injured spouse. However, a psychiatrist must show unconditional remorse and compassion towards all partners, even the partner that has committed the betrayal, if treatment is to be effective, and the infidelity is not to recur. These are the rugged shoals that might ruin marital therapy after infidelity. On the one side, often, the psychiatrist must take the role of the unfaithful partner so that he/she doesn't believe that an infinite sequence of emotional beatings would be counseling. On the other side, without being forced to believe like she/he was the source of the infidelity, the deceived partner must hear the motives for her/his partner being unfaithful.

## Tips for Effective Communication

Keep these ideas on good negotiation skills in mind every time you're grappling with confrontation, and you will produce a more optimistic impact. Ok, here's how.

### Stay Focused

Even while coping with current ones, it's easy to dig up previous apparently connected disputes. It seems necessary to fix all that concerns you at once to have it all discussed whilst you are still struggling with one disagreement. Unfortunately, this also clouds the dilemma and allows it less possible to achieve common consensus and a remedy to the present problem, and allows the whole conversation more taxing and even confusing. Try not to dig up hurts or other problems in the past. Keep concentrated on the moment, the thoughts, empathy, and seeking a remedy for each other. Practicing meditation on mindfulness will enable you in all aspects of your life to strive to be more aware.

**Listen Carefully**

People sometimes believe they are listening, but when the other person is talking, they are just worrying about what they are going to discuss next. Remember to remember that the next time you do it, you're in a debate. Good contact really goes in both directions. Although it may be challenging, strive to genuinely listen to what your partner is doing. And don't disturb it. Do not get protective. Only listen to them and think back on what they mean because they realize that you've understood it. Then you will have a greater view of them, and they will be more able to listen to you.

**Try to See Their Point of View**

All of us primarily want to be heard and acknowledged in a confrontation. To bring the other individual to see it our way, we speak a tone from our point of view. This is understandable, but too much of an emphasis will backfire on our own ability to be heard above all else. If we all do this all the time, ironically, there's no emphasis on the point of view of the other guy, and no one feels heard. Try to always see the other side, and maybe you can describe yours more. (If you don't understand it, ask more questions before you do.) If they feel understood, others are more inclined to be able to respond.

## Respond to Criticism with Empathy

It's quick to believe like they're mistaken and get defensive anytime someone comes at you with criticism. While feedback is painful to hear and sometimes exaggerated or colored by the emotions of the other individual, it is necessary to listen to the distress of the other individual and react with respect for their feelings. Look at what is real about what they mean, too; at you, it can be useful knowledge.

## Take control of What's Yours

Realize that power is a personal duty, not a vulnerability. Good contact requires knowing that you're mistaken. If both of you bear the blame in a dispute (which is generally the case), search at what is yours and confess to it. This diffuses the case, sets a good precedent, and reflects sophistication. It very always motivates the other person to behave in a kind, bringing you all closer to shared understanding and a solution.

## Use "I" Messages

It's less argumentative, sparks less condescension, and lets the other party consider your point of view instead of feeling threatened, rather than doing something like, "You really fucked up here," start sentences with "I," and making them about yourself and your emotions, such as, "I feel upset when this occurs."

## Look for Compromise

Look for options that satisfy the interests of all instead of seeking to "win" the case. This emphasis is far more productive than one party having what they want at the detriment of the other, either by compromise or a new innovative approach that offers both of you what you want most. Good contact means seeking a settlement in which all parties will be satisfied.

## Take a Time-Out

Tempers get hot occasionally, and it's just so hard to maintain a conversation without it being a disagreement or a battle. It's safe to take a break from the conversation before you all cool down if you sense yourself or your spouse beginning to get too upset to be productive or exhibiting any destructive communication habits. This can include taking a stroll and calming down in half an hour to return to the discussion, "sleeping on it" so that as long as you return to the discussion, you can process what you feel a bit better or whatever seems like the right match between the two of you. Effective contact also requires understanding when to take a rest.

**Keep at It**

It is also a safe thing to take a break from the debate and then come back to it. If you also handle the issue with a positive mindset, shared interest, and a desire to consider the other's point of view or at least pursue a compromise, you will make strides toward the aim of a resolution to the dispute. Don't give up on contact until it's time to give up on the partnership.

**Ask for Help If You Need It**

If either or both of you have difficulty being polite during the confrontation, or if you have attempted to settle confrontation on your own with your spouse, and the problem just doesn't seem to change, you might profit from a few therapy sessions. Counseling for spouses or family counseling can assist with altercations and teach skills to solve potential conflicts. You will also profit from traveling solo if your partner doesn't want to go.

## After the Infidelity: Can Counseling Help?

It may be difficult to negotiate the depths of marital therapy when one spouse has become unfaithful. Historically speaking, holding a mindset of unconditional good regard towards both the betrayer and the deceived is the trickiest aspect. Very predictably, the psychiatrist tends towards showing the strongest sympathy towards the injured spouse. However, a psychiatrist must show unconditional remorse and compassion towards all partners, even the partner that has committed the betrayal, if treatment is to be effective, and the infidelity is not to recur. These are the rugged shoals that might ruin marital therapy after infidelity. On the one side, often, the psychiatrist must take the role of the unfaithful partner so that he/she doesn't believe that an infinite sequence of emotional beatings would be counseling. On the other side, without being forced to believe like she/he was the source of the infidelity, the deceived partner must hear the motives for her/his partner being unfaithful. Concern about their kids is the most typical explanation about partners to see psychiatrists when a spouse has had an affair. The hurt spouse will continue to remain in the marriage, particularly when the children are small, if she/he really loves her / his spouse, and believes that he/she genuinely repents and that there will be no potential infidelity. He/she still has to assume this if a psychiatrist is to support the pair have a happier life. Each spouse is often advised to see an independent psychiatrist, who will unconditionally take their hand. As a marital psychologist, it is important that a psychiatrist is on the side of the marriage/relationship going forward, and all spouses may be encouraged to voice their counseling needs.

# What Helps the Couple Heal?

## The Frame of PTSD

It helps to understand that with recurrent hallucinations and a desire to go over and over the specifics of the infidelity in counseling, the deceived partner struggles from a type of post-traumatic stress disorder (PTSD). Going over and over incidents throughout the era in which the scandal took place provides a sense of power to the hurt partner. For the betrayer, who usually needs to move forward with the marriage and put the past behind them, this desire to continually rehash the specifics of adultery can be challenging. The betrayer does not comprehend the need of his spouse to go constantly in counseling for the same terrain, especially when the specifics give her too much distress. Framing the negative emotions of the injured person as a form of PTSD allows the betrayer to gain more empathy with the method. In as much depth as she wants and as often as she wants, he must address all of his spouse's concerns. About even the slightest detail, he must be absolutely truthful.

## Saying "sorry" every day

The injured partner may continue to hear the partner who has been unfaithful apologies for the infidelity every day in the early stages of counseling. During the week, she will even like him to email her or call her more regularly so that she knows that her husband often cares about her. When the hurt spouse has noticed repeated messages or calls between her husband and the entity with whom he had an affair, this is extremely important.

## A ritual of burying the past

A practice of symbolically placing infidelity behind them is strongly recommended at the middle stage of the counseling. It is therefore advised that any reminders of the time of the adultery be buried by the pair. The burial is expected to take place somewhere dear to them and sacred. This sort of tradition serves to drawback happy feelings of the moment when the pair courted and cherished the most.

## The need to feel sexually desirable

The hurt spouse also wants to be told that she/he is as physically attractive as the other individual. In counseling sessions, it is necessary to provide a frank dialogue of what each spouse requires to encourage desire.

## The end of therapy

If relationship counseling has been effective, the conversation between the pair can transition from continually rehashing the particulars of infidelity to addressing more daily forms of marital concerns, such as budgets, privacy, how they invest their time, and so on. The injured partner has recovered a certain degree of faith at this point, and the betrayer has learned to communicate his emotions and desires through counseling. While it can be frustrating and challenging to counsel a couple during infidelity — for the therapist as well as the spouses; it can also be successful. A survey showed that following infidelity, 71 percent of partners a psychiatrist had met in counseling were together.

## Build a Healthy Relationship Based on Interdependence

Most of us, especially in our intimate relationships, respect interactions with others. In reality, we are wired for communication, and it helps us to establish a bond with our spouse and to

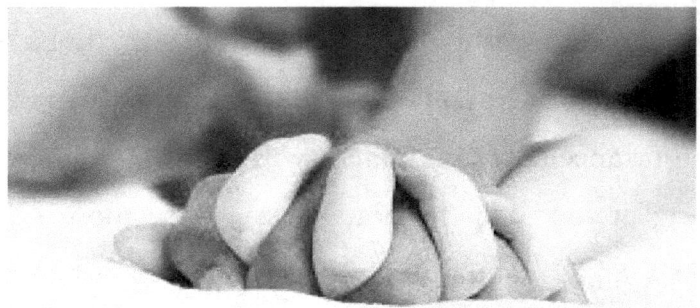

establish intimacy. Long-term relationship performance relies heavily on the nature of our intimate relationship with each other.

When we think of our perfect relationships, we always think of our most significant partner as a beautiful, close, lifetime partnership. How can we create a partnership of that kind? That cozy, secure, long-lasting bond with someone we know has our long-lasting back? A relationship that allows us the freedom to be ourselves, that facilitates our progress and enables us to be flexible with each other? Understanding the distinction between interdependence and codependence is one of the main factors.

## What Is Interdependence?

Interdependence implies that the strength of the relational bond they share is understood and respected by spouses while retaining a clear sense of self throughout the relational relationship. It may sound frightening or even unhealthy to be reliant on another person. Growing up, we are always given over-inflated importance of individuality, with a strong value put on not having anyone for emotional help, to be more self-contained. As vital as maintaining a sense of control is taken to an extreme, this may potentially get in the way of us being able to communicate in a positive way with others emotionally. For those who have an exceptional sense of individuality, intimate interaction with a spouse may be challenging to attain, even

terrifying or not perceived as especially valuable in a relationship.

**Interdependence Is Not Codependence**

The same thing as being codependent is not interdependence.
A codependent human, for their sense of self and well-being,
appears to rely heavily on others. There is little opportunity for
the entity to discern where they stop, and their spouse starts.
There is an enmeshed sense of obligation to another person to
satisfy their needs and/or to feel okay with who they are with
their spouse to satisfy all of their needs.

- A codependent relationship's traits contain items such
  as:

- Poor / no thresholds

- People-pleasing attitudes

- Tolerance

- Unhealthy, inefficient contact

- Tampering

- Emotional relationship challenges

- Behavior management

- Blaming one another

- Poor self-esteem of either of the couples or both

- No individual agendas or objectives outside the relationship

Codependent experiences are not safe and do not allow space for spouses to be themselves, to evolve, and to be independent. Either spouse or both depend strongly on the other, and the relationship for their sense of confidence, feelings of worthiness, and general mental well-being are included in these dysfunctional relationships. For either or both parties, there are always emotions of remorse and embarrassment when the relationship isn't going on.

**Why Interdependence Is Healthy for a Relationship**

Interdependence requires a compromise between oneself and others within the family, understanding that all parties are functioning in acceptable and substantive ways to be involved and fulfill the physical and emotional needs of each other.

Partners are not intrusive from each other and do not search for emotions of worth to their partner. In times of need and the ability to make certain choices without fear of what will happen in the relationship, this allows each spouse space to retain a sense of independence, room to step towards each other.

# How to Build an Interdependent Relationship

Being mindful of who you are from the outset is the secret to creating an interdependent relationship. People sometimes search for or join into relationships merely to stop becoming lonely, without any personal reflection on who they are, what they admire, and their relationship priorities. Taking time for this sort of personal contemplation encourages you to enter into a new relationship with a self-awareness that is essential for an interdependent relationship to be formed. The trick to maintaining a safe, interdependent relationship would be to give your spouse space and the chance to do this same stuff. Starting the relationship in this way will enable both parties to build a comfortable room to learn how to transform intimately towards each other without fear of sacrificing themselves or being influenced or exploited. Interdependent marriages do not leave individuals with their spouse or the relationship feeling guilty or afraid, but rather leave them feeling comfortable with their spouse. Taking time in the most meaningful relationships to focus on who you are and what you desire. In the dating phase, being aware of this will help guarantee that partnership can remain healthier and better in the long run.

# Chapter 6: Formula for Building a Healthy relationship

We slip into a relationship too much and hang onto it for way too long. An addictive relationship, where there is a cluster of very positive stuff that each person clings to in place of chronic tension and the breakup and get back together loop, is a big indication of incompatibility. The pair sometimes goes further into denying their incompatibility when such a relationship does not succeed and battle it by delving into their relationship anymore; by moving in together, getting hitched, getting married, raising a son, constructing a home, moving and beginning over, etc. Other incompatible persons, afraid of being lonely and/or afraid of losing stability, are tolerably happy with each other, so they hang on like lifeless statues. Emotional intimacy can be described as enabling yourself, by acts that convey emotions, weaknesses, and trust, to communicate more closely with your spouse. Sharing your secrets, dreaming about your life, and asking your vital partner news is part of a relationship. When both sides will express and appreciate each other's emotions, a pair is usually happy.

## The Formula for Staying Together

Divorce has always left us fearful. Maybe it began by seeing the grief of your near relative and the deep wounds remaining from the breakup. Researchers have provided scientific evidence for decades and provide statistical formulas for what functions and what does not work. We are mindful that there must be at least five pleasant experiences (an embrace, a true love looks, sincerely pleasant comments) with any negative encounter a pair has (eye roll, hostile body expression, real unpleasant utterance, etc.).

Notice that the term genuine has been repeated above twice. However, a common mandate suggests "fake it before you make it" in cognitive-behavioral counseling. Our nonverbal signs sometimes betray how we feel. If you utter flattering terms and do a bunch of pleasant things but don't really feel it or believe it, odds are you're adding up a tone of negative nonverbal experiences. You can mix a hundred negative nonverbal realities in one fake nice thing you do.

**This is a huge takeaway**

At the end of a relationship, how many of us lament, "I tried too hard"? Was it true, or was it focused on a combination of anxiety, remorse, wanting to fit in, obeying command, grandiosity, dominance, protection, or sex? Just something to remember, and while you are pondering that, here are a couple of the risky experiences that may present the bringer of breakups. Other supportive evidence shows that the harbinger of breakups could be presented.

## Criticism

Yup, it's terrible. Give it as a recommendation, not a critique, when we want to make our partner dress smarter. And there's undoubtedly a message here in respecting them as they are and not caring about what people say in order not to betray how you actually feel about their outfit or etiquette in public. Unfortunately, at the very beginning of a partnership, we have all been guilty of this. Huh. Yikes. Criticism, though, can get uglier than reporting about a mismatched dress and public belching. It might get mean. It expands into the next level of tragedy when it's violent enough.

## Contempt

Displeasure and contemptuous comments hurt in a war. They are more than words that not only ruin relationships; they break away the members of the surrounding family and the wellbeing of everyone. Oftentimes, when a couple is experiencing the same old fight, disdain falls in. That a script should be published is too chronic. The cause-varies, but the combat stays the same. A fundamental gap in personality or other unmet essential needs lies beneath the war, which takes us to the next reaction.

## Defensiveness

If we don't get what we really want or need out of this relationship, and we don't feel comfortable, valued, and protected anymore, so before they can say or do something, we already start preparing our answer to our spouse. Interestingly, evidence reveals that men rehearse distress-maintaining thoughts rather than women. Such conditions in the midst of a Texas summer drought contribute to the worst of all experiences that practically scorch love like an untended vine.

## Stonewalling

It's a fighting pair, and one steps away. A bit of a time-out is okay, but if there is no repair, and each participant continues to respond entirely to the thoughts and viewpoint of the other party, a literal wall is created. The pair will eventually be disengaged and so deeply apart that they will not communicate at all. Why does partnership contribute to all of this? Ok, we know that with every bad one, we require five good experiences. We all realize that we should not falsify optimistic relationships, and our body language and nonverbal signs can expose our fact, and when we fight, we can actually go to the limit and create a few (or all) of the contact bombs just mentioned. Here are some of the options.

## Know Yourself

Try getting to know yourself and your values and fundamental needs until you can even connect to another partner. The pressures imposed on your wife can often be alleviated by concentrating on you. You find that through yourself and relatives, partners, hobbies, and spiritual development, you will fill the unmet needs. On most occasions, it shows a deep divide for your partner to hear about you and your beliefs.

## Check Compatibility

We slip into a relationship too much and hang onto it for way too long. An addictive relationship, where there is a cluster of very positive stuff that each person clings to in place of chronic tension and the breakup and get back together loop, is a big indication of incompatibility. The pair sometimes goes further into denying their incompatibility when such a relationship does not succeed and battle it by delving into their relationship anymore; by moving in together, getting hitched, getting married, raising a son, constructing a home, moving and beginning over, etc. Other incompatible persons, afraid of being lonely and/or afraid of losing stability, are tolerably happy with each other, so they hang on like lifeless statues.

## Signs of Compatibility

- Shared trust.

- Allowing the point of view of each person to be understood and considered (mutual influence).

- Consideration for the growth, disagreements, and limits of each other.

- Good time separately and together.

- Emotional bonding.

- True joy together.

- Five or more good against one negative.

- Spiritual relation and honoring the views of each other.

- Relevant alliance.

- Actually knowing each other and how each other feels and thinks.

- Fighting equally and having equal styles of war and repair.

- Volunteering and learn about each other.

- Compatible principles and priorities of life.

- Romantic chemistry.

- Involvement.

## Achieving Emotional Intimacy

The cornerstone of any strong partnership is emotional intimacy. To strengthen the relation, you have with your partner, here are a few items you can do. In an intimate relationship, as we mention affection, what typically comes to mind are physical actions, such as holding hands, cuddling, hugging, and even sex. Although physical intimacy is vital to any romantic bond, it is one of the key reasons that separates it from any other form of relationship; it is just as necessary to cultivate emotional intimacy, if not more.

### What is Emotional Intimacy? And Why does it matter?

Emotional intimacy can be described as enabling yourself, by acts that convey emotions, weaknesses, and trust, to communicate more closely with your spouse. Sharing your secrets, dreaming about your life, and asking your vital partner news is part of a relationship. When both sides will express and appreciate each other's emotions, a pair is usually happy.

Ultimately, inside the relationship, the relational trust provides

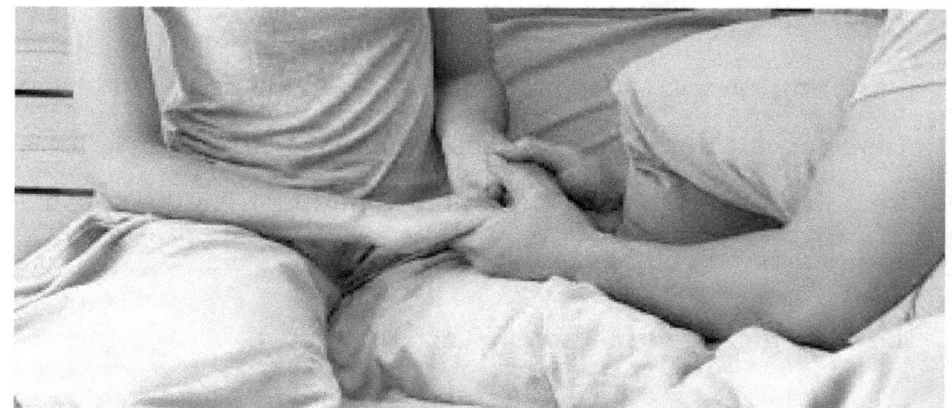

a profound sense of confidence and a willingness to be completely yourself, quirks and all, without feeling as though you are compromising the relationship itself. A partnership fails in several respects without this intimacy. You may feel resentful or bitter, experience oversensitivity, have fears about the commitment of your spouse to you, or experience feelings of alienation or loneliness, for example. If there is a loss of emotional intimacy, either or both of you will experience a lack of protection, affection, encouragement, overall attachment, and in an intimate relationship, it would therefore most definitely impact physical intimacy. Getting an intimate relationship without personal intimacy is not viable in the long run. When you think of emotional intimacy as the cornerstone of every relationship, spending your capital (time, money, and energy) in creating it and continuing to cultivate, it always becomes a no-brainer.

## Ways to immediately improve Intimacy

Fostering interpersonal intimacy is an evolving process that can require some time to learn, as many items. There are a few items you should do, though, beginning tonight, strengthening the intimate bond you have with your wife.

### Be strategically vulnerable to earn their trust.

Even after we've invested a massive amount of time with somebody, it's hard to knock down our personal barriers often. Although you can't cause someone to become weak, you can render yourself weak by going out of your way. Critically relevant is the practice of strategic weakness. Choose one place to begin instead of attempting to be insecure in any aspect of your life. This might translate into discussing something that occurred at work that you would not have addressed otherwise, voicing a thought you had that was tough to convey in the past, or disclosing a reality you've been hanging on to about yourself.

### Give your partner daily affirmations and compliments.

If you're in a relationship for six months or 60 years, it's easy to take the good qualities of our spouse for granted and often tough to communicate how deeply we cherish them. Having a practice of offering your spouse specific compliments and affirmations can help you retain a viewpoint on why this person is important to you, and it can help them realize that you see them. You never want to feel invisible to your partner when you have failed to express your gratitude. Such verbal comments may be as plain as stating, "I want you to realize how much I respect you" or "I really appreciate the time it took you to do x, y, or z."

**Prioritize sexual satisfaction**

Research showed that when they were sexually fulfilled, partners recorded having a stronger emotional bond. The two are inextricably related in that way. Although getting sex alone is not a cure-all to strengthen the relational relationship, taking the time to understand and discover the needs of the partner; and getting the same reciprocated; will result in stronger relational attachment feelings in and out of the bedroom.

**Make an effort to break out of your day-to-day routine.**

It's possible to reach a comfort zone plateau with how busy life is, in which we brush past each other, simply attempting to cross things off our to-do lists. This is in stark contrast to the start of a partnership, where it feels fresh and thrilling to do everything we do, even where we go beyond and above. This may imply that we have lost sight of the importance of doing something for each other that in the other individual produces pleasure or affection. We stop trying to please, we stop trying to grasp, and insecurity and emotions will get lost in the regular routine in such settings. It is extremely necessary that we find time for each other in a more meaningful way than either getting dinner together or going to bed together. Garner motivation in a partnership from those early courting days. If you schedule a spontaneous square dance date night for beginners, you decide to go for ice cream and a walk; you turn up with roses "only because" or you sit down and plan a weekend getaway together.

## Factors in Healthy Relationships

Intimate relationships, particularly at the very beginning, can go from seamlessly easy to extremely complicated in what seems to be the blink of an eye. Although, if people are frank about themselves, this twitch of an eye is also more like a prolonged amount of time because, when issues arise, they keep their eyes closed or avert their eyes. Then, it may be impossible to find out what happened and what to do to remedy it by the time they actually look at it. Couples, instead, will do better to aim specifically at their concerns. They will then analyze and find a way to solve certain challenges. Since a two-person situation needs a two-person solution, the only way to tackle this is together. And the more both individuals are involved in nurturing their relationship, the greater the potential for a solid, romantic bond that relationship provides. In addition to only resolving conflicts, the other aspect of maintaining a relationship healthy is dreaming of what will make things happier. These needs, of course, that partners consider what a good relationship looks like. At that end, ten important factors for a stable relationship are presented below.

- Liking and loving each other sincerely.

- Doing stuff mostly to make one another happier.

- Enjoying and valorizing time together and working diligently to make it possible.

- A capacity to express and acknowledge the love

- A deep sense of relationship commitment; a desire to hold to the relationship despite momentary disagreements and moments of disinterest or even hate.

- Good skills for teamwork and problem-solving

- A commitment to function peacefully across tensions and differences, coupled with the capacity to forgive and embrace forgiveness.

- Realistic and settled upon shared goals; with a desire to measure up to certain goals

- A common life ideology, namely principles, and goals. This is very thorough and very critical. It entails, for example, mutual beliefs towards family and partners and a common parenting style (for those with children).

- Sexual relationship is fulfilling

If your partnership is faltering, or if you simply want to change it, think of each of these variables. Chat with your partner about them. Decide, and focus on which areas may use enhancement. Consider these concerns when you plan to communicate with your partner: to what degree does your relationship have any of the variables above? Where are you going to improve? And, in your relationship, what other considerations do you think are important? In order to be genuinely stable, we noticed three main elements that relationships must-have. It's not just about the dishes, the trash, or even the income, as too many couples believe it is, while couples are arguing with each other, and it's one of those blood-boiling sorts of arguments. If relational bonds are not safe, and couples feel isolated from each other, every form of partnership offers appropriate grounds for a war. However, the substance is not what the war is about. The main topic of "are you there with me?" is what they are actually fighting about.

**Accessibility**

Accessibility is the first main element of safe interactions.

People ought to feel as though they are open to their partners, and their partners should be available. Pay attention to your partner and be alert to when it seems they are attempting to contact you in order to improve accessibility in your relationship. In periods of disconnection, it may always be challenging to offer an olive branch, but your partner can attempt to approach you after a battle, but in a gentle way. Only aim to be open to this. In order to only listen, it is also necessary to be accessible. People only want to be understood by their partners too many times, and they long for empathy, but they get an unwelcome solution. By just listening and validating how your spouse thinks, you will maximize your accessibility. To be validated, still feels amazing.

**Responsiveness**

In stable marriages, the second main element is responsiveness. It will seem clear this one. React when your partner comes up to you. If you are, in reality, unable while you are doing something else, let them realize that their issues are important to you and encourage them. Find a later time where you will come together and address the topic and honor the dedication, in fact. When couples tend to freeze each other out and do not react to each other, they open up all sorts of troublesome possibilities for their relationship. By answering, instead, stay linked.

### Emotional Engagement

In stable marriages, the third main element is a personal dedi-cation. Emotions have not always been well known; however, the further analysis contributes to an improved perception of them. Rather than anything else, love is just an emotional bond, and studies in psychiatry, psychology, and biology appear to back up this argument. Therefore, it is important that couples are personally connected with each other. Not only is it neces-sary to think about the emotional experience of your spouse and be concerned about it, but you should let them know as well. The more couples are physically connected with each other, the deeper their relationship is. The next time you and your wife get into one of those blood-boiling battles, pause, take a deep breath and remind yourself what you're actually battling for. Chances are, you both fail to see whether just how much you truly mean to each other if you are there supporting each other. Increase the responsiveness, reactivity, and emotional in-teraction with each other, and it will continue to be simpler to resolve fights, when they will always be all about the dishes, the trash, and the money, of course.

## How to Make Relationship Therapy Effective

Successful counseling relies not just on the counselor's talents and expertise but also on the clients' ability. In order to make marital therapy more successful, there are several items you should do.

## Be Honest

Do not lie to the psychologist. We often lie because we do not want to be judged. The role of your psychiatrist, though, is not to assess you but to support you. Even when it's rough, remain real.

## Prepare Yourself for Discomfort

Therapy will also create anxiety when you are learning fresh, not all of them good or happy, realities about yourself. Working on yourself allows you to calm back and recognize in your discomfort that you need to develop and change. Your therapist is there to support so really it up to you to do the job.

## Listen to Your partner.

It's crucial to listen to what people have to suggest whether you are doing relationship counseling for one person or a wider family community. It's just going to make things more complicated for you to sit on the defensive and attempt to react to something people bring up regarding your actions.

## Put in the Time

Therapy occurs in sessions almost as often as it does in them. In between visits, your psychologist can send you homework or encourage you to pursue different communication and relationship habits. It can take time and commitment, but keep in mind that it is worth it. In the end, it's the role placed by all participants of the relationship that creates a difference in the therapy's outcomes.

**If Your Partner Refuses Therapy**

Even if you feel that counseling will help your partners, your wife may not be able to take part. But, in this case, what should you do? It's crucial to note that your partner should not be coerced into therapy. Ultimately, what you can do is see a psychiatrist of your own and focus on problems that you experience as a person.

## Myths About Couples Therapy

Therapy for partners may be overwhelming, and some ideas do not improve. What would your response be if anyone were to inquire how you felt about going to a couple's therapy? Perhaps the thought will make you feel impatient, excited, or relieved. If so, you will surely not be the first person to feel that way. But it would also make a lot of sense if you were to catch yourself smiling at the premise, or feeling uneasy, indecisive, humiliated, or reluctant, to name only a few answers. The thought of going to counseling for partners may be overwhelming, and there are different explanations for this. For one, counseling for partners requires vulnerability. Imagine sitting alongside your wife side by side while you all recount to a stranger the most intimate aspects of your life and partnership.

**Couples Therapy gets a Bad Reputation.**

The grim mental image connected with it is another factor. As the location where marriages go to die, couples therapy gets a poor name, but making a move to reach out to a couple's counseling may feel like an acceptance that the relationship must take its last breaths. For certain individuals, couple counseling may even bring up the idea of a lion's nest, with at least one person becoming cautious about going because they want their person or the therapist (or both) to perceive them in the relationship as the only obstacle and making them "the bad guy."

**Sense of Inadequacy and Shame**

In addition, with those partners who criticize themselves for being unwilling to overcome a dilemma on their own, there is a sensation of inadequacy and guilt that may emerge. When speaking about couple counseling, other people experience a feeling of futility since their failure to alter the complexities of their relationship seems like a confirmation that couple counseling will not really improve. All and all, it is quickly known that people are not in a huge rush to pursue counseling for partners. Here, we'll dig at some of the theories about counseling for partners and attempt to unpack them.

"For persons to go to pair counseling, breakup or to split up may be on the agenda."

### Multiple Reasons for going to Therapy

There are diverse explanations for partners to go to counseling. True, after consulting with a psychiatrist, certain couples are on the brink of separating. In reality, there are couples that can function in a successful way with a psychiatrist to end their relationship. One or both parties feel uncertain if they want the partnership to proceed like other couples in counseling. This was the case with around 20 percent of the couples in one sample. Fourteen percent of couples began counseling in another survey to find out whether they should salvage the relationship or whether they could part ways. There are legitimate, substantive motives for pursuing the care of partners.

- For multiple but equally legitimate purposes, a number of other couples' step in the entrance gate of couple's therapy and start working on their relationship. One team of researchers, for instance, found that about 46 percent of couples wanted to cope more with disagreements, 30 percent wanted to repair their relationship, and 25 percent of couples thought they were doing good and only wanted to strengthen another aspect of their marriage.

- Similarly, another group of researchers noticed that 57 percent of couples wished to enhance their closeness or emotions to each other, and the same number aimed to change how they relate and talk. In comparison, 32 percent wanted to cope with their children's concerns, 28 percent were trying to strengthen their physical interaction, 15 percent were inspired by their partner's love, and 10 percent wanted to concentrate on preserving their relationship's strengths. What this study shows us is that there are not only multiple motives for families to undergo counseling, but there are also persons who see it as a means to improve their partnership and be the best partners they can be. In other terms, marriages may not appear to be in extreme trouble or have major difficulties for spouses to choose to try the care of couples.

"If I go to counseling for couples, I'll only get accused and physically assaulted. No thanks at all."

**Still, there is no 100% success rate.**

Although we can't literally promise that this won't happen, we can assure you that it isn't meant to happen. It will not allow all partners in the room to have a sense of protection. It's crucial for couples, not just for their own good, but for the good of their relationship, to feel like they have a warm therapeutic connection with their clinician. Data shows that this indicates a change in their intimate relationship as partners have a better bond with their therapist. In other terms, while the psychiatrist does not always agree with you and does require improvements from you and the partner, you should feel understood and encouraged, not verbally assaulted or treated in the relationship as the villain.

"We shouldn't have to go to counseling with couples. We should be able to resolve things by ourselves."

### Asking for Help is a Sign of Strength

A victory of courage is the choice to call for support. Instead of marching on and going it alone, it can be hard to reach out, particularly for couples who believe they are supposed to be able to change what's troubling them without any outside support. The concern with the assumption that partners are expected to fix relationship problems on their own is that it does not map to how individual relationships work. Not all relationship dynamics, for instance, are plain and simple, particularly for the people within the relationship, because often what appears to be the issue is not the real problem. It's kind of like grappling with an unseen boxer. How are spouses expected to accurately fix the situation because, by no fault of their own, they can't see it? A therapist for a partner, someone who provides an outside perception of the relationship, may help partners perceive their relationship from a particular viewpoint and connect in a new way.

**It doesn't map onto human needs**

Another problem with this definition is that it does not map all forms of treatment that we get from individuals. How many of us have found support from a psychologist, a nutritionist, a personal trainer, or a career mentor without blinking? We don't really agree that we should be able to work out, diet right, develop our jobs, or take control of our wellbeing on our own. Do marriages have to be any different?

**Why go to couple's therapy? It won't work**

A heap of doubt on whether pair counseling will actually succeed makes sense because couples have tried and tried to change their relationship, and little has succeeded. And there's no way to guarantee a relationship can change injustice. This is part of the hazard. Things do not get easier. Around the same time, perhaps because they have not come across what fits, not because their circumstance is irreparable, the explanation couples can't understand how their relationship might improve. And there are techniques that run, such as Emotionally Oriented Counseling (EFT), which has solid clinical evidence support. So, while it is true that counseling for families will not work, it may inevitably surprise partners with how much it may make a difference. Of course, it's entirely right to decide that counseling for spouses just doesn't seem persuasive, meaningful, or significant. But for those couples who may attempt but don't, we hope that because of some of the challenges we've talked about, this can provide a little support in coping with the disheartening challenges that spouses often see in their way.

# Conclusion

It is so popular to claim "relationships are complicated" that it's a cliché now. Yet it's real as well. Stress and everyday life can create disputes that appear complicated or sometimes unlikely to overcome, even though individuals get along pretty well. Relationship therapy may assist persons in progressing with their issues in these stressful circumstances, stepping through them, and be happier partners overall. We feel comforted and encouraged, our anxiety is minimized, and we learn that our attachment figures can be relied on in difficult times when our attachment figures react to our anxiety in ways that fulfill our needs. But if parents frequently respond to the frustration of a child by downplaying their feelings, ignoring their requests for support, or having the child feel stupid, the child will learn not to accept their attachment figures for support and to ignore and cope with their problems and feelings alone. When dreaming about getting into a long-term, serious relationship, passion is a top priority. In reality, 88% of Americans say that the most important factor to consider getting married is passion. We want to be cherished by our partner and fall in love with her. More than ever before, marriages are met with more strain. Couples are often dealing with relational communication difficulties and holding love intact, in addition to the long-standing stress of topics such as finances, life changes, and family dynamics. When we do not experience that kind of bond in our relationship, we turn to our spouses for warmth, reassurance, and closeness and feel hurt. Partners can get trapped in dysfunctional cycles of disconnection and start worrying with the time that they are no longer supposed to be together. "Couples counseling" and "couples therapy" are usually considered as the same thing. On a scientific basis, there is little distinction between them. The other way in which whether the session is called matters is a valid one; in certain places, you may get a separate "therapy" qualification or license to practice that is tougher to receive than the "counseling" qualification or license to practice. This form of relationship with a trained therapist presents partners with an ability to work on their most complicated or socially demanding concerns, whether you name it partners therapy or couples counseling. These topics may vary from basic difficulties of understanding or serious disputes to difficulties of drug misuse and psychiatric conditions. Although counseling for partners may be a wonderful way to bond with your spouse or mend the gaps between you, without having a therapist, there are also ways to ensure sure you maintain the flame is going and the relationship safe. There are several tools out there that rely on couples counseling ideas or tests. The cornerstone of any strong partnership is emotional

144

intimacy. To strengthen the relation, you have with your partner, here are a few items you can do. In an intimate relationship, as we mention affection, what typically comes to mind are physical actions, such as holding hands, cuddling, hugging, and even sex. Although physical intimacy is vital to any romantic bond, it is one of the key reasons that separates it from any other form of relationship; it is just as necessary to cultivate emotional intimacy, if not more. We know life is hard, and relationships are hard. Loneliness is hard. But we have to choose what kind of "hard" we want. We need to be aware and address the issues while they are starting out before they become a much larger problem for our peace of mind as well as our relationships.

CPSIA information can be obtained
at www.ICGtesting.com
Printed in the USA
LVHW021032291220
675197LV00002B/405